Control Systems for Air Conditioning and Refrigeration

Control Systems for Air Conditioning and Refrigeration

BILLY C. LANGLEY

014531

Prentice-Hall, Inc.
Englewood Cliffs, N.J. 07632

Library of Congress Cataloging in Publication Data

Langley, Billy C., (date)
   Control systems for air conditioning and refrigeration.

   Includes index.
   1. Air conditioning—Control. 2. Refrigeration and refrigerating machinery—Automatic control. I. Title.
TH7687.5.L364 1985   697.9'32   84-16086
ISBN 0-13-171679-4

Editorial/production supervision and
   interior design: **Nancy Milnamow** and **Tracey L. Orbine**
Cover design: **Diane Saxe**
Manufacturing buyer: **Anthony Caruso**

© 1985 by Prentice-Hall, Inc., Englewood Cliffs, New Jersey 07632

All rights reserved. No part of this book may be
reproduced, in any form or by any means,
without permission in writing from the publisher.

Printed in the United States of America

10 9 8 7 6 5 4 3 2

ISBN 0-13-171679-01

Prentice-Hall International, Inc., *London*
Prentice-Hall of Australia Pty. Limited, *Sydney*
Editora Prentice-Hall do Brasil, Ltda., *Rio de Janeiro*
Prentice-Hall Canada Inc., *Toronto*
Prentice-Hall Hispanoamericana, S.A., *Mexico*
Prentice-Hall of India Private Limited, *New Delhi*
Prentice-Hall of Japan, Inc., *Tokyo*
Prentice-Hall of Southeast Asia Pte. Ltd., *Singapore*
Whitehall Books Limited, *Wellington, New Zealand*

In memory of our dear son, Billy C. Langley, Jr.

Contents

PREFACE   xv

1   TYPES OF CONTROL SYSTEMS   1

    Control System Components   4
        The control process   4
        Disturbance sensing element   5
        Controller   5
        Final control element   6

    A Simple Control System Example   6

    Control Modes   7
        On-off (two-position) control   7
        Multiposition (multistage) control   8
        Floating control   8
        Proportioning (modulating)) control   9

    Complex Variations of the Proportioning Control   10

    Review Questions   10

2 BASIC CONTROL THEORY — 12

Controlled Systems Characteristics and Elements   12
    Controlled system   12
    Controlled variable   13
    Manipulated variable   13

Control Equipment   13
    Set point   13
    Control point   13
    Desired value   13
    Deviation   13
    Corrective action   14
    Differential gap   14
    Proportional band   14
    Cycling   14
    Offset   14
    Lag   14
    Primary element   14
    Final control element   14

An Automatic Control System and the Basic Functions of Its Parts   15
    Controllers   15
    Temperature sensing primary elements   16
    Pressure sensing primary elements   17
    Humidity sensing primary elements   17
    Controller mechanisms   17
    Methods of transmitting energy to the actuator   18
    Actuators   18

Modes of Automatic Control   18
    Two-position control   18
    Simple two-position control   19
    Timed two-position control   20
    Proportional control   22
    Floating control   23
    Proportional-plus-reset control   23

Lag   23

Review Questions   24

Contents									ix

**3	BASIC PNEUMATIC CONTROL SYSTEM					26**

    Advantages of Pneumatic Control Systems   29

    Review Questions   29

**4	PNEUMATIC CONTROL SYSTEM COMPONENTS			31**

    Pneumatic Controllers   31
        Direct-acting controller   31
        Reverse-acting controller   32

    Single-Pressure Thermostat   32

    Basic Dual-Pressure Pneumatic System   36
        Summer/winter system   36
        Day/night system   39

    Temperature Controllers   41

    Pressure Controllers   44

    Master/Submaster Controllers   46

    Receiver Controllers and Transmission Systems   49

    Review Questions   57

**5	PNEUMATIC CONTROL VALVES						59**

    Valve Flow Characteristics   63
        Quick opening   63
        Linear   63
        Equal percentage   63

Valve Ratings and Terminology 63
    Capacity index 63
    Close off 64
    Close-off rating for three-way valves 64
    Maximum fluid pressure and temperature 64
    Pressure drop 64
    Critical pressure drop 64
    Body rating 64
    Nominal body rating 65
    Rangeability 65

Close-Off Ratings and Spring Ranges 65

Steam Valves 66

Supply and Return Main Pressures 67

Design Considerations for Steam Coils 67

Steam to Hot Water Converters 68

Steam Humidifiers 69

Selecting Steam Valves 70

Water Valves 73

Water Circulating Systems 77

Supply Pressure Differential Regulation 78

Water Valve Selection 79

Review Questions 82

6   PNEUMATIC RELAYS   83

Diverting Relays 83

Reversing Relays 84

Pneumatic-Electric Relays 85

Contents   xi

    Electric-Pneumatic Relays   86

    Pressure Selector Relays   89

    Air Motion Relays   90

    Averaging Relays   91

    Pneumatic Switches   91
        Gradual Switch   92
        Minimum positioning switch (accumulator)   92

    Review Questions   93

7   FUNDAMENTALS OF ELECTRONIC CONTROL SYSTEMS   95

    Basic Electricity   95
        Ohm's law   95
        Series connections   96
        Parallel connections   97
        ac and dc power   98
        Power in dc circuits   100
        Power rating of equipment   102
        Inductance   102
        Capacitance   102
        Power in ac circuits   103
        Semiconductors   105
        Rectifiers (diodes)   105
        Zener diodes   106
        Silicon controlled rectifiers   107
        Transistors   108
        Bridge theory   109

    Sensors   110
        Temperature elements   110
        Humidity elements   111

    Controllers   112
        Single-element controllers   113
        Dual-element controllers   116

Actuators   117
    Electrohydraulic actuators   117
    Thermal actuators   118

Auxiliary Devices   119

Review Questions   120

8   BASIC ELECTRIC CONTROL CIRCUITS   121

Definitions   121
    Line voltage   122
    Low voltage   122
    Potentiometer   122
    Balancing relay   122
    In contacts and out contacts   122

Classification of Electric Control Circuits   123
    Series 40 control application   125
    Series 40 control equipment   126
    Series 40 control operation   126
    Series 40 control combinations   127
    Series 80 control application   128
    Series 80 control equipment   129
    Series 80 control operation   129
    Series 80 control combinations   130
    Series 10 control application   131
    Series 10 control equipment   131
    Series 10 control operation   132
    Series 10 control combinations   135
    Series 60 floating control application   136
    Series 60 floating control equipment   138
    Series 60 floating control operation   139
    Series 60 floating control combinations   141
    Series 20 control application   141
    Series 20 control equipment   142
    Series 20 control operation   143
    Series 20 control combinations   147
    Series 60 two-position control application   151
    Series 60 two-position equipment   151
    Series 60 two-position control operation   152
    Series 60 two-position control combinations   152

Contents

Series 90 control application   152
Series 90 control equipment   153
Series 90 control operation   154
Series 90 control combinations   158

Review Questions   169

INDEX   **171**

Preface

The purpose of this manual is to provide information on the operation and application of pneumatic, electronic, and electric control systems.

The concepts presented here apply regardless of any changes in the concepts and design of control systems for air conditioning and refrigeration units.

Every effort has been made to relate the principles and the actual application of controls; however, this is not a conclusive or an exhaustive study of the constantly changing methods used in designing control systems for air conditioning and refrigeration units.

The thorough understanding of this manual will give the reader the knowledge and confidence necessary to service and install control systems related to the air conditioning and refrigeration industry.

Billy C. Langley

Control Systems
for Air Conditioning
and Refrigeration

1

Types of Control Systems

Control systems are generally classified according to their source of power. When we consider controls from this point of view, the components which go into the makeup of a control system are either electric, self-contained, or pneumatic control devices, as follows:

1. *Electric*. The source of power for these controls is electricity. The control units in this system are connected to the source of power by electric wires which may carry either line-voltage or low-voltage electrical current. The line voltage, when used, may be either 120 or 240 V. The low-voltage control circuit normally operates on 24 V, which is supplied by a transformer. The transformer is used to convert the 120 V or the 240 V to 24 V.

A line-voltage thermostat which operates a heating control motor to open a hot water, or a steam, valve and close it when required by the room temperature is an illustration of a simple application of a line-voltage electric control. (see Figure 1-1).

A simple application of a low-voltage electric control would be a motor operating on a low voltage which is supplied by a transformer (see Figure 1-2). Except for the low-voltage transformer, the operation is exactly like the line-voltage circuit presented in Figure 1-1.

2. *Self-Contained*. These controls have as a source of power the sensing element, and the final control mechanism—such as a valve—is all contained in one single unit. These types of controls usually include a bellows and a bulb which are connected together by a length of tubing containing a charge of liquid or vapor which changes its volume with a change in room temperature. Thus, a change in the temperature of the medium surrounding the sensing element will cause the

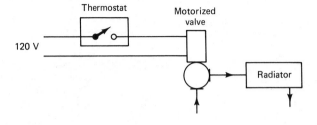

Figure 1-1 Line-voltage two-position control.

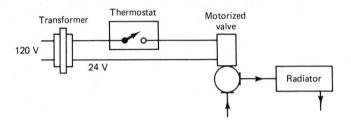

Figure 1-2 Low-voltage two-position control.

fluid inside the sensing element to either expand or contract with a sufficient amount of force to operate the valve.

A simple application of the self-contained control is illustrated in Figure 1-3. The valve which controls the amount of hot water or steam which enters the heating chamber is directly operated by these changes in the pressure inside the bellows as the temperature of the surrounding medium changes. The operating principle of these controls is very similar to the operation of a pneumatic control system. The basic difference is that the variations in pressure at the bellows are produced in direct response to the variations in temperature at the sensing element.

3. *Pneumatic.* When a pneumatic control system is employed, the control system components are connected together by air lines and respond to air pressure which is supplied by an air compressor. This air is usually stored in a tank with the pressure about 100 psig. The air is supplied to the control system at a pressure of about 15–25 psig through a pressure regulator. These air lines may be made of either copper or polyethylene and vary in size from $\frac{1}{8}$ to 2 in. in diameter. The most common sizes are from $\frac{3}{16}$ to $\frac{1}{2}$ in. in diameter.

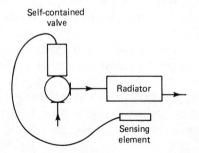

Figure 1-3 Self-contained control valve.

Chap. 1   Types of Control Systems

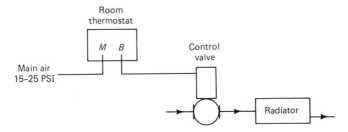

Figure 1-4 Simple pneumatic control.

A simple application of a pneumatic control system is shown in Figure 1-4. In operation, an operating pressure of 15–25 psig is supplied to the controller through the main air line from the compressor. The thermostat is in effect a pressure regulator which is actuated by the surrounding air temperature.

The line which supplies air to the radiator valve motor maintains a pressure of about 15 psig or less, in response to the room temperature, and is known as the branch line. The valve shown is a normally open type. It will open as the air pressure decreases. As the thermostat responds to a fall in the room temperature, the air pressure is decreased in the branch line, allowing the valve to open and pass steam to the radiator. A rise in temperature at the thermostat causes an increase in branch line air pressure, causing the pneumatic motor to close the valve and stop the flow of steam to the radiator.

The control valve motor, which consists of a bellows which is balanced against a spring, does not waste air. However, when the valve motor is changing positions, air does not flow into or out of the motor. The valve assumes a position which is determined by the air pressure in the branch line, the tension of the spring, and the force imposed on the valve stem by the steam pressure. Therefore the thermostat regulates the air pressure and effectively determines the position of both the motor and the valve.

The conditioned space in any heating, ventilating, or air conditioning installation may be controlled in many different ways, such as the following: as a single unit, with the building divided into different zones, or with each room controlled individually. The single unit is the most popular type of control system in use today and is most popular in residential air conditioning installations. As the size of the building is increased, the zone type of control system becomes more popular because it is more difficult to obtain proper control for the whole building with a single thermostat mounted on the wall. Thus, the building is split into separate zones, each having its own individual thermostat and control system. The individual room control systems are by far the most accurate and satisfactory for any type of building, whether it is a residence or a commercial building. These types of control systems have a separate thermostat located in each room, and it has control of the heating and the cooling equipment. The thermostat in each room controls the equipment in response to the requirements of that room regardless of the conditions in another room. This type of control system is usually very expensive because of the number of devices required throughout the entire building. However, the cost may not be a determining factor when flexibility and a very accurate type of control system are required to obtain the most satisfactory results.

CONTROL SYSTEM COMPONENTS

In an effort to simplify the control system, it can be represented by use of a block diagram. See Figure 1-5. Each of the blocks in this diagram represents an essential component which is used in a control system.

The Control Process

The control process is probably the major requirement of any control system. When there is nothing to control, we have no need for a control system. A person who walks away from a wrecked automobile with the steering wheel still in his/her hand has a vital control element, but there is no system to be controlled. Likewise, a very elaborate control system which is used to control a steam heating system is not a system if the steam is removed from the pipes. The process can be in the form of the maintenance of a required temperature in a hospital operating room or the operation of a steam boiler.

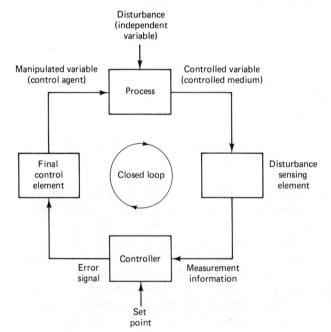

Figure 1-5 Block diagram of a simple control system.

Disturbance Sensing Element

The conditioning process is subject to some type of external influence, known as the disturbance, such as a change in the set point, the supply, the demand, or the environment. If it were not for these needs, there would not be any need for the control system. The disturbance factor itself is considered an independent variable which cannot be changed by the system. It may be compared to an army officer who issues orders to his subordinates, who in turn must make any required adjustments in their actions in order to follow the officer's commands.

Any disturbance present will cause a change in the controlled variable. The controlled variable is the quantity or the condition which is being controlled. The controlled variable is therefore a characteristic of the controlled medium, which is the substance being controlled. *Example*: When controlling the temperature of water, the controlled variable is the water temperature, and the controlled medium is the water itself.

For the disturbance to be counteracted, it must first be detected and measured. The control system must know how much change is required in the controlled variable to counteract the disturbance. This detection and measuring are the responsibility of the disturbance sensing element. It is not possible for a human to determine the pressure which exists inside a pipe by looking at the outside of the pipe or to determine the temperature of a flame inside a firebox by looking at the color of it. There are transducers, fortunately, of many types which will obtain this type of information for the control system. Some of the more common types of disturbance sensing elements that provide measurement information are pressure sensing, temperature sensing, and humidity sensing.

Controller

The controller is the device that receives the measuring information from the disturbance sensing element and interprets it to determine how the process is going. The disturbance sensing element, in many cases, will be a part of the controller mechanism in the form of a bimetal as in a thermostat. A set point, or a reference standard, is established in the controller mechanism. When the conditions match the set point, the process is doing well, but when they do not match, the controller produces an error signal which will initiate the required corrective action.

In some cases the set point may be manually set, as for the temperature setting of a thermostat, and in such a case it will remain at this setting for a long period of time. In some cases it may be adjusted automatically, as in a boiler system where the steam requirements are referenced to the input and the airflow is adjusted according to the demands. In this case the reference standard is constantly varying.

Final Control Element

The final control element is a mechanism which changes the value of the manipulated variable in response to an error signal from the controller. The manipulated variable is a characteristic of the control agent, which causes the desired change in the controlled variable. *Example*: steam which is flowing through a coil to heat a room; the steam is the control agent, and the amount of flow through the coil is the manipulated variable. The flow of steam is varied in response to any changes in the room temperature, the controlled variable, the air, or the controlled medium in the room or space.

The final control element is the device which causes the manipulated variable and the control agent to make a correcting action on the process operation. In this case, the control system is an error-sensitive, self-correcting system, usually referred to as a closed-loop feedback control system.

A SIMPLE CONTROL SYSTEM EXAMPLE

As an illustration of the components in a control system, we may compare the hardware of a simple water-level controller (see Figure 1-6) to the block diagram shown in Figure 1-5. The purpose of the process is to maintain a specific amount of water in the tank. The water level is the controlled variable. The water in the tank is the controlled medium. A disturbance is present when the position of the output valve is changed to vary the amount of water flowing into the tank. This change in water level is detected by the float, the disturbance sensing element. The measurement information is then transmitted to the float valve linkage, which in this case becomes the controller. The water-level height is determined by adjusting the adjustment on the float valve linkage, an adjustment of the set point. The float linkage, the controller, then transmits the error signal to the input valve,

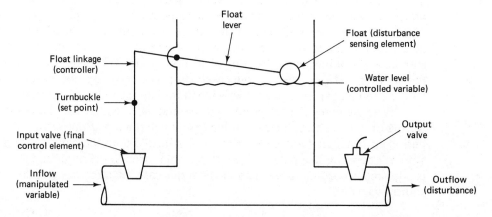

Figure 1-6 Simple control system.

Chap. 1   Types of Control Systems   7

the final control element. The volume of water flow, the manipulated variable, is thus changed by the repositioning of the input valve to make the desired correction in the water level, the controlled variable. In this particular system, the water itself is the control agent and the controlled medium.

CONTROL MODES

A control mode is the method that a control system uses to make the desired corrections in response to a disturbance. It makes the proper relation between the operation of the final control element and the measurement information which is given by the disturbance sensing element. It is generally, therefore, a function of the controller itself. The proper matching of the mode to the desired process is the determining factor for the overall performance of the control system. There are four basic modes: (1) on-off (two-position), (2) multiposition (multistage), (3) floating, and (4) proportioning (modulating). Some of the more complex variations of the proportioning mode may include the addition of the reset action and rate action or in some cases both.

On-Off (Two-Position) Control

The on-off type of control, as its name suggests, provides only for a two-position control operation. This type of system is either full on or full off. There are no intermediate positions available in the operation. When there is a predetermined amount of variation in the controlled variable from the set point, the final control element will move to either of the extreme positions which are available. The amount of time between the on and off positions will vary in accordance to the demands from the load. See Figure 1-7.

The on-off type of control system is the simplest mode. It also has some definite disadvantages in that it will allow the controlled variable to vary over a range rather than settling down to an almost steady condition. This range is sometimes quite wide. If this range is set to be too narrow, the controller will become worn out by the constant switching on and off action.

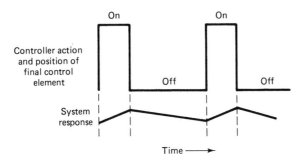

Figure 1-7 On-off (two-position) mode of system control.

Multiposition (Multistage) Control

Multiposition (multistage) control is an extension of the on-off type of control system where the system is equipped with two or more stage functions. In cases where the range between the on and off positions is too wide to provide the desired operation, multiple stages which have much smaller ranges are used. Each of these independent stages has only the on and off operation, but there are many positions which are available by simply adding more switches. This action results in a steplike operation. The more stages that are used, the smoother the operation will be. On an increase in the load, more stages are turned on to handle the additional requirements. In most cases, the multiposition control will provide from 2 to 10 operating stages. See Figure 1-8.

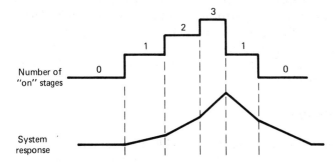

Figure 1-8 Multiposition mode of system control.

Floating Control

Floating control is different from the on-off type or the multiposition type in that the final control element can assume any position between its extremes. The controller proper has two positions with a neutral zone between the two. When the controller is in the neutral zone, the final control element will not move from this position. When the disturbance becomes large enough, the controller will move to one of its two positions. This action causes the final control element to move in one direction or the other in response to the position of the controller. The final control element makes this change at a constant speed, and therefore this mode is commonly called a single-speed floating control. As the controller moves to the neutral zone, the final control element will stop. It will remain in this position until the controller again moves to one of its two positions. This mode is therefore called a floating mode because the final control element stops when the controller is floating between its two positions. See Figure 1-9.

The floating type of control system is used in applications which require gradual changes in the load requirements, or where there is a small lag between the disturbance and its detection. The final control element speed should be fast enough so that it can keep pace with the more rapid load changes. If its speed will not provide for this action, hunting and excessive cycling will result.

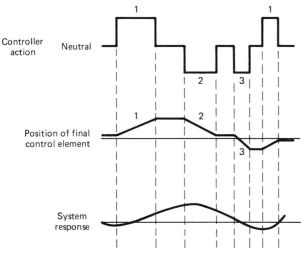

1 AND 2 indicate position of controller

Figure 1-9 Floating mode of system control.

Proportioning (Modulating) Control

In a proportioning (modulating) control system, as in the floating type, the final control element can assume any position between the two positions of the controller. Such systems are different from floating control in that they have no neutral zone. Therefore, a disturbance, no matter how small, will cause the controller to move. Every movement of the controller has a specified amount of movement for the final control element. These movements are made as often as the disturbances occur. The proportioning type of control produces a linear relation between the amount of disturbance and the amount of controller action and the position of the final control element. See Figure 1-10.

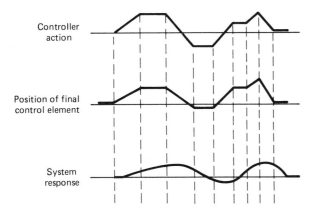

Figure 1-10 Proportioning (modulating) mode of system control.

COMPLEX VARIATIONS OF THE PROPORTIONING CONTROL

When a large change in the load occurs, an offset may occur between the value of the controlled variable and the set point. A constant adjustment of the set point is required to maintain the controlled variable at the same set point through the complete load range of the equipment. Sophisticated types of controllers make use of two types of automatic action which are frequently added to ensure the best possible control under the most severe conditions that may occur during the process.

When the reset action is used, the error signal from the controller is varied in proportion to the amount of offset and the amount of time that it is present. The final control element will continue to move in a direction which will correct the error. It will stop only when the error signal becomes zero; at this time the controlled variable is at the set point.

When rate action is used, the error signal from the controller is varied in proportion to the rate at which the disturbance occurs. It is used for two purposes: (1) to accelerate (speed up) the return of the controlled variable to the set point in a part of the process which is slow in responding and (2) to anticipate a disturbance which may occur and start the corrective action before any change is detected.

REVIEW QUESTIONS

1. What are the three types of components that make up a control system?
2. In a pneumatic control system, what air pressure is supplied to the system?
3. What are the three major types of control systems?
4. What is the most popular type of control system for residential units?
5. What is a change in the control set point of a controller known as?
6. What control system component receives and interprets the measuring information?
7. What control system component changes the value of the manipulated variable in response to an error signal from the controller?
8. Name the four basic control modes.
9. What type of control system is best suited to gradual load changes?
10. In the proportional type of control system, describe the reset action.
11. Why are large systems more adaptable to zone control?
12. What is probably the major requirement of any control system?
13. What must happen for a disturbance to be counteracted?
14. What causes the desired change in the controlled variable?
15. To what type of system is the error-sensitive, self-correcting system referred?
16. What type of equipment operation does the two-position control provide?

Chap. 1   Types of Control Systems   11

17. What type of control system provides the simplest mode of operation?
18. What type of control is an extension of the on-off type of control system?
19. What two types of controls can assume any position between the full open and the full closed positions of the controller?
20. In what two cases is the rate action used?

2

Basic Control Theory

Automatic control systems and their applications vary in the range from simple domestic temperature regulation to the precision control of complex industrial processes. Therefore, automatic control systems can be used in any place that a variable condition needs to be controlled. The variable condition may be a solid, a liquid, or a vapor, and it may be in the form of temperature control, pressure control, the control of humidity, or the rate of a flow.

The most important consideration in controlling these conditions is in the operation of the controlled and the controlling devices.

CONTROLLED SYSTEMS CHARACTERISTICS AND ELEMENTS

For an automatic control system to bring about its effects, there must be a variable that needs to be controlled. Controlling this variable is often accomplished by an automatic control system which controls the second variable. The second variable is called the *manipulated variable*. The manipulated variable is what causes the required changes in the controlled variable.

Controlled System

A controlled system contains all of the equipment in which the controlled variable exists, but it does not include the automatic control equipment.

Controlled Variable

The controlled variable is that which is actually controlled and measured. The controlled variable is in the controlled medium. *Example:* When the temperature of water is being controlled, the controlled condition is the temperature, and the water is the controlled medium.

Manipulated Variable

The manipulated variable is that quantity of the condition which is being regulated by the automatic control system and causes the desired change to happen in the controlled variable. The manipulated variable is a characteristic of the control agent. As an example, assume that we have a heat exchanger in a gas-fired furnace which is used to heat a room. The room thermostat is in a position to measure the temperature (controlled variable) of the room air (controlled medium) and operates a gas valve which regulates the flow of gas (manipulated variable) into the combustion chamber. The heat from the heat exchanger is used to heat the room air.

CONTROL EQUIPMENT

There are several other terms that the technician must become familiar with before an effective discussion of automatic control systems can be discussed in any detail.

Set Point

The set point is the position on the controller at which the controller indicator is set.

Control Point

The control point is the point of the controlled variable which the controller operates to maintain.

Desired Value

The desired value is the value of the controlled variable which it is desired to maintain.

Deviation

The deviation is the difference between the set point and the value of the controlled variable at any instant in time.

Corrective Action

The corrective action brings about a change in the manipulated variable and is initiated by a deviation.

Differential Gap

The differential gap is a term which is applied to two-position controllers. It is the smallest range through which the controlled variable must pass while moving the final control element from one position to the other in a two-position controller. It is also called *differential*.

Proportional Band

The proportional band is the point of a positional positioning controller through which the controlled variable must pass when moving the final control element through its full operating range. It may also be known as *throttling range* and *modulating range*.

Cycling

Cycling is a periodic change in the controlled variable from one point in the process to another. It is also called *hunting*.

Offset

Offset is a sustained deviation between the point of the controlled variable which corresponds to the set point and the control point.

Lag

Lag is the amount of delay in the effect of a changed condition at one point in the control system to some other condition to which it may be related.

Primary Element

The primary element is that portion of the controller which uses any energy derived from the controlled medium first to cause a condition which represents the point of the controlled variable. As an example, a bimetal in a thermostat is a primary element.

Final Control Element

The final control element is that part of the controller mechanism which directly acts to cause a change in the value of the manipulated variable.

Chap. 2   Basic Control Theory   15

AN AUTOMATIC CONTROL SYSTEM AND THE BASIC FUNCTIONS OF ITS PARTS

For a control to be fully automatic, there are six basic functions that it must perform, as follows:

Function 1: measuring the changes in one or more of the controlled conditions or variables. *Performed by:* the sensing and measuring element of the proper controller.

Function 2: translating those changes into forces or energy forms that can be used by the final control element. *Performed by:* the controller mechanism.

Function 3: the transmitting of the energy or forces from the point of translation to the point of the corrective action. *Performed by:* the connecting members of the particular control circuit: the wiring for electric circuits, the piping for pneumatic systems, and the linkages for mechanical equipment.

Function 4: the use of the force or energy to position the final control element and to effect the proper corrective change in the controlled condition. *Performed by:* the controlled device, such as a valve or a motor.

Function 5: the detection of the completion of the corrective change. *Performed by:* the sensing and measuring elements of the particular controller.

Function 6: the termination of the call for corrective action and the prevention of overcorrection of the condition. *Performed by:* the controller mechanism, connecting means, and the actuator or the controlled device.

For these functions to be performed, an interconnection between the required number of controllers is necessary. In most cases, the type of energy which is best suited for the type of control problem which is being considered will determine the kind of equipment which is selected for the job. The most common forms of energy available for control systems are compressed air and electricity. Because of this, our discussion in this book will be focused on pneumatic, electric, and electronic controls and systems.

Controllers

A controller may be defined as a device which does two things: (1) It senses and measures any changes in the controlled variable, and (2) it uses an impulse which is received due to the sensing and measuring of the controlled variable to meter the energy which is of a form usable in the control circuit. This metered energy then actuates the control equipment, which will then correct some change or prevent any further change in the variable being controlled.

The duties of sensing and measuring are performed by the primary element in the controller. The type of material and construction of the primary element must allow the primary element to respond to any changes in the controlled condition. Pneumatic and electric controls use basically the same kinds of primary elements. Electronic controls, because of their extreme sensitivity, are capable of using primary elements of a type to which pneumatic and electric controls are not capable of responding.

Temperature Sensing Primary Elements

The primary element in a primary controller is usually made of either a bimetal strip, a sealed bellows, or a sealed bellows attached to a remote sensing bulb.

A bimetal strip is a thin material made from two different types of metal. One of these metals will expand at a faster rate than the other, causing the curvature of the metal to change with a change in the temperature. Because of this movement, the bimetal can be used to open or close a set of contacts which control the flow of current through an electrical or electronic circuit or to regulate the flow of air in pneumatic control circuits.

This bimetal strip is sometimes wound into a spiral shape to allow a greater length to be used in a much smaller space. Other applications use a simple U-curved-type bimetal.

Sealed bellows are also used as primary elements which are designed to sense temperature. In manufacture, the bellows is evacuated and is partially or completely filled with a fluid, usually a liquid. As the temperature changes, the bellows will expand or contract as the pressure inside the bellows changes and causes the controller mechanism to move in the desired direction. When a remote bulb type of thermostat is used, a bellows to which a remote bulb or capsule is attached by a length of capillary tubing is used. The remote sensing bulb is placed in the controlled medium. Any changes in the temperature of the controlled medium will also cause changes in the pressure of the fluid in the sensing bulb assembly. These changes in pressure are transmitted through the capillary tubing to the bellows located inside the thermostat. When the bellows expand or contract, the thermostat mechanism is caused to move in the proper direction.

Electronic temperature controllers, because of their sensitivity, are capable of using primary elements which are made from a much lighter material which will respond quickly to even small changes in the temperature of the controlled condition. The signal put out by the primary element is relatively small, but the electronic circuit can easily magnify, or amplify it to a usable strength.

The primary element which is used in an electronic circuit may be only a wire whose resistance changes with a change in temperature. When used in flame controllers, the variation of the ionization in the flame itself can be used by these devices. An alternate method is to use the photoelectric cell which is placed where it can sense the flame.

Pressure Sensing Primary Elements

Controllers with pressure sensing primary elements use bellows, diaphragms, inverted bells immersed in oil, or other similar pressure-sensitive devices to accomplish their task. The medium which is under pressure may transmit its pressure directly to the controlling device, and the movement caused by this pressure may be used to operate the mechanism of an electric or pneumatic controller. The rate of flow, quantity of flow, liquid level, and static pressure are all some type of variation of the pressure control principle.

Humidity Sensing Primary Elements

The primary element which is used in a humidity controller is usually made from human hair, leather, wood, or any other type of substance which will expand and contract with the changes in humidity. As the humidity in the surrounding area changes, the moisture is either absorbed or released, causing the element to expand or contract, operating the controller mechanism.

The humidity sensing element which is suitable for use on an electronic controller is made from two comblike strips of gold fused to a piece of glass. Electrical contact is made between the two strips by a hygroscopic salt which has been painted on the surface of the glass. The resistance of the salt varies with the amount of moisture it has absorbed. This small change in the resistance causes a change in the current which is detected and put to use by the electronic circuit.

Controller Mechanisms

The next thing to be done to accomplish automatic control is to translate the measured change of the controlled variable into some form of energy which can be used by the control system in doing its job. The primary control element is the device which transforms the measurement of the controlled condition into a usable impulse. The impulse then acts upon the controller mechanism and causes it to react in the prescribed manner.

When an electric controller is used, the impulse from the primary element is used to open or close an electric circuit or to set up a varying resistance in an already established circuit. When pneumatic controllers are used, the mechanism is already made from a system of valves, which are opened or closed, or an air vane, which is used to regulate the air pressure to the final control element by bleeding off air into the atmosphere.

The controller mechanism is, in effect, an amplifier. This is especially so when electronic circuits are used, because the sensing and measuring procedures are carried out in terms of electronic energy and are amplified by an electronic amplifier.

Methods of Transmitting Energy to the Actuator

When electric and electronic circuits are used, wiring is used to transfer the energy from the controller to the actuator. When pneumatic control systems are used, copper tubing or plastic piping is used for this purpose. Diagrams are used for both the electric and the pneumatic control systems to show the sequence of connection and operation desired for the system being considered.

Actuators

An actuator is a controlled motor, relay, or solenoid which converts the electric, electronic, or pneumatic energy into rotary, linear, or a switching action as desired. The actuator can cause a change in the controlled variable by operating a number of kinds of final control elements, such as valves and dampers.

There are many different types of control motors or actuators which operate on both electric and pneumatic systems. Pneumatic motors are usually of the modulating or proportioning type, and thus they can assume a position at any point in their moving range depending on the air pressure which is delivered to their mechanism. Relays must be used with these controllers to obtain a definite on or off action.

Electric motors are either two-position, floating, or proportional positioning. Some of them are unidirectional and rotate through the complete 360°, and other types are limited in their stroke and have two directions of travel.

MODES OF AUTOMATIC CONTROL

All types of automatic control systems do not use the same types of control action to accomplish their desired purpose. The method used is called the *control mode*. The most common types are the two-position, floating, proportional-position, proportional-plus-reset, and the different variations of these controls.

All modes of control except the floating type and those which have an automatic reset have the inherent characteristics of offset. Offset is a sustained deviation between the value of the controlled variable which corresponds to the control point and the set point. If the offset were maintained at the same value at all times and at all load conditions, it could be changed by a simple calibration of the specific controller. However, this point varies with the load conditions. Even though the amount of offset can be predicted when the characteristics of the particular application are known, the equipment which is required to make the necessary change is usually very expensive and is often complicated.

Two-Position Control

When two-position controls are used, the final control element occupies one of the two possible positions, except for the brief period of time when the control is passing from one of these positions to the other. The two-position type of control

Chap. 2   Basic Control Theory

is popular on applications such as controlling the off and on operation of simple heating systems, refrigeration systems, air conditioning systems, spray ponds, and humidification equipment and for energizing and de-energizing electric heat strips.

Two-position controls respond to two values of the controlled variable when determining the position of the final control element. The zone between these two values is known as the *differential.* The controller cannot initiate any type of action of the final control element in this zone. When the controlled variable approaches the upper of these two variables, the final control element assumes one of the two positions corresponding to the demands of that specific controller. The final control element will remain in this position until the controlled variable returns to the lower value. At this point, the final control element will move to the other position immediately and remain there until the controlled variable again reaches the upper variable.

There are currently two types of two-position controllers. The first to be developed was the *simple two-position control.* This type of control has been, more or less, the industry standard in the past. A later development is the *timed two-position control,* which is rapidly replacing the simple two-position control.

Simple Two-Position Control

In this type of control, the controller and the final control element interact with each other in the same manner as those without modification from any source, either mechanical or thermal. This operation results in a cyclical operation of the equipment and a condition during which the controlled variable cycles back and forth between the two variable points which are determined by the differential of the controller and the inherent lag in the system. Because the controller action is such that it cannot change the position of the final control element until the controlled variable has reached one of the two limits of the differential, these limits then also become the minimum possible swing of the controlled variable. See Figure 2-1.

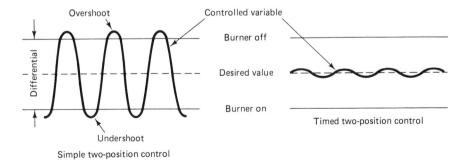

Figure 2-1 Curves comparing variations in the controlled variables for simple two-position and timed two-position control.

When a simple two-position control is used, the controller never reaches the controlled condition. Therefore, it corrects a condition which has already passed rather than correcting one that is happening or one that is about to happen. Because of this fact, the simple two-position control systems are applicable only to those systems which have a small system lag. Thus, the simple two-position types of control systems are seldom used in comfort conditioning control systems. They do, however, lend themselves quite well to some types of industrial processes and some types of auxiliary processes used in air conditioning applications.

There is not a single control point in a simple two-position control system. Thus, the controlled variable cycles between the two extreme control points. It is easier to think of the control point as being at a point halfway between the two extremes and the offset as being a point which is a sustained deviation of this control point. The offset is, therefore, a shifting of the complete cyclic curve, either up or down, and the mean is either raised or lowered so that it is no longer at a point midway between the upper and the lower limits of the controller differential.

The offset is caused by the fact that the average value of the controlled variable must be at a lower value when it is subjected to heavy load conditions and higher when the load conditions are lighter in order that the heat can be supplied at a higher or lower rate than needed. When the system is operating at a peak load, the burner must operate 100% of the time. Because of this, the controlled variable cannot raise the upper limit of the thermostat differential or the burner would be turned off. Also, under minimum load conditions the controlled variable cannot fall to the lower limit of the thermostat differential or the burner would be put into operation.

The amount of controller offset is in this way limited by the controller differential. Therefore, it is not usually a serious problem in the simple two-position-type control unless it happens that a wide controller differential is needed.

Timed Two-Position Control

The preferred method of heating any space or building is to replace the heat in the exact amount that is lost as it is lost. This cannot be accomplished with the two-position type of control because the equipment is either on or off and the amount of heat supplied to the space is either too much or too little. However, we can get very close to the desired amount of heat delivery by using a timed two-position type of control system. When using this method of heat control, the heat is delivered on a *percentage on-time control*, and the fluctuations of the control point are virtually eliminated.

As an example, let's assume that we have a residential heating unit using a two-position type of control which is required to make up a heat loss of 60,000 Btu in 1 hr at a given load condition. The total capacity of the equipment is 120,000 Btu/hr. This condition requires that the equipment operate 30 min. out of this particular hour. It may be required to operate on any desired combination of on and off times to provide the required amount of heat to the space.

A large combination of long operating cycles would probably be unsatisfactory because of the wide variations in the temperature. By dividing the heat delivery into the proper amounts at the proper times, a closer control of the heat is obtained with greater satisfaction.

When timed two-position control is used, the basic interaction between the controller and the final control element is the same as it is for the simple two-position control system, with the exception that the controller responds to the gradual changes in the average value of the controlled variable rather than to the cyclical fluctuations of the simple two-position controller. These gradual changes modify the timing action in accordance with the load.

The desired timing action may be provided by some type of mechanical means such as a cam action. The greatest disadvantage of this method is that only the relative duration of the on and off periods may be varied as the load conditions change; the frequency of operation remains fixed.

The thermal timing devices are much more flexible and convenient than the mechanical timing devices. These devices use a heating element and a sensing element placed close to each other in a single housing. Controlling the electrical power to the heating element creates a thermal timer. As long as the ambient temperature remains within certain limits, the thermal timer will cool to its on point, energize the heater, heat to its off point and de-energize the heater, and again cool to its on point, and the cycle is repeated. With a change in the ambient temperature the time required for the timer to operate is also changed. A decrease will cause the timer to take a longer period of time to start the equipment. Therefore, the timer automatically changes the ratio of on time to off time. Also, the nonlinear shape of the heating and cooling curves may be used to vary the total cycling time of the equipment and thus the frequency of the cycles.

In electronic control systems the ambient temperature at the timer is effectively held constant by an ambient temperature compensator. The on and off points are automatically reset by remote temperature elements such as a room thermostat, an outdoor compensator, or any other type of suitable temperature controller or controllers. A raising of the operating range of the cycling control is equivalent to a reduction in the outdoor temperature which causes an increase in the amount of on time.

When any type of timed two-position control is used, it offers a great advantage over the simple two-position type of system in that it greatly reduces the swings in the controlled variable which are the results of a large total lag. See Figure 2-1. Because the controller does not need to wait for a signal from the controlled variable to detect the cyclic changes and signal for some type of corrective action, the control system lags have no significant effect on system operation. The inherent lags in the equipment and the distribution system serve only to help smooth out the variations in the heat delivery so that the results closely approximate those of a continuous delivery system equipped with a proportional-position control system.

A factor in offset in a timed two-position control is the addition of heat to the thermostat bimetal. When the cycle of the thermostat used in this type of control is analyzed, it may be seen that the control point must vary if the bimetal is to cool and heat at the different rates necessary to time the cycle for the various load conditions. With a decrease in the outside air temperature the heat which has been lost from the space also increases, and the on cycle of the equipment must also be increased to replace the lost heat at the proper rate. Thus, the heating rate of the thermostat bimetal must be slower to allow the burner to operate for a longer period of time. The cooling rate of the bimetal during the off cycle must be faster so that the equipment will be brought on sooner. Both of these conditions require that the difference between the bimetal temperature and the air temperature becomes greater. This temperature difference is maintained by a deviation of the room temperature, known as the offset.

In a timed two-position control system, the offset is equal to the total heat which is added minus the manual differential of the thermostat. The total heat is equal to the difference between the maximum temperature of the bimetal and the temperature of the air surrounding it. The manual differential is the differential for which the thermostat is set.

Proportional Control

In proportional-type controls, the final control element will move to a position from the set point which is proportional to the deviation of the value of the controlled variable. There is only one position which is attainable by the final control element for each value of the controlled variable occurring within the proportional band of the controller used. Because of this, the position of the final control element appears as a continuous linear form equal to the value of the controlled variable.

Due to the fact that there is only one position attained by the final control element for each of the values of the controlled variable, a sustained deviation of the control is necessary in order to place the final control element in any position other than the midposition of its operating range, assuming that the set point is in the midposition of the proportional band. The offset, then, becomes a major problem in the proportional-position type of control device.

For example, we have a proportional control directing a hot water coil which is used to heat a residence. When ideal load conditions exist, the thermostat is in the midposition of its proportional band, the coil valve will be half open, and there will be no offset. Now let's suppose that the outside air temperature drops, increasing the load on the hot water coil. Almost instantly, more heat is required to replace the heat which has been lost from the room at a greater rate than before the outside air temperature dropped. To deliver the required amount of additional heat, the water valve must open further and remain open until the load has been reduced. To accomplish this feat, the temperature must deviate from the set point

Chap. 2   Basic Control Theory

and sustain the required deviation because the position of the final control element is proportional to the amount of the deviation.

On an increase in load from the ideal point, the offset will increase toward a colder position; also, as the load decreases from the ideal, the offset increases toward the warmer position.

Floating Control

The floating type of control is a mode of control in which the final control element moves, at a predetermined rate, in the corrective direction until the controller itself is satisfied or until a movement in the opposite direction is required by the controller. The direction of this movement will correspond to the direction of the deviation of the controlled variable. The floating type of control may be further divided into several subclasses. The following are two of these subclasses which are of interest to us: *Single-speed floating control* is one in which the final control element is moved at the single speed throughout its entire range of effectiveness. *Proportional speed control* is one in which the final control element is moved at a rate which is proportional to the deviation of the controlled variable.

Either of these types of control systems is very adaptable to systems which have a fast reaction rate, a slight transfer lag, and a slow change in the load. Generally, the porportional speed control may be used in systems which have a somewhat faster change in the load conditions than those which operate successfully with a single-speed floating control.

Proportional-Plus-Reset Control

In proportional-plus-reset control systems, both the proportional control and the proportional speed control may be combined so that they act together to position the final control element. Thus, any given response from the final control element will have two components. The component used in the proportional speed floating control is termed *reset response.* The component used in the proportional control is termed *proportional response.* The reset response has a tendency to correct the offset resulting when the proportional control is being used alone. The effectiveness of the resetting action is what determines the remaining offset.

LAG

Lag is the amount of delay in a response from one part of a system to a changed condition in another part of the system. The total lag is the sum of the individual lags in the control system and the lags in the system which is being controlled.

The lag due to controller lag consists of time lags in the following:

1. The controller. This is predominantely a measuring lag.

2. The final control element.

The lag due to system lag consists of time lags in the following:

1. A change in the manipulated variable in response to any changes in the position of the final control element
2. A new rate of energy being transferred from the control agent and the controlled medium
3. A response of the controlled variable to changes in the energy which is transferred to it from the control agent

When a temperature controller is used, it is easy to see that a lag could exist because the transfer of heat is not an instantaneous process. As an example, a thermostat is controlling the temperature of a room to within the specified limits, and a fairly stable condition is the result. If a large amount of cold air were introduced into the room in a short period of time, the room temperature would soon drop. A lag would result from this condition because the sensing element of the thermostat could not sense this temperature drop immediately because it would take a few seconds for the sensing element to cool to the new temperature. It should also be obvious that a lag will occur between the release of the heat into the room and the room being warmed to a point which will satisfy the thermostat.

The lags which occur in pressure control systems are generally smaller than those in temperature control systems because the pressures can be sensed, measured, transmitted, and corrected at a much faster rate.

REVIEW QUESTIONS

1. When are control systems used?
2. Define the controlled variable.
3. To what is the manipulated variable characteristic?
4. What is the position on the controller at which the control indicator is set called?
5. What is the point of the controlled variable which the controller operates to maintain?
6. Define deviation.
7. To what type of controller does the term differential gap apply?
8. What is another name for throttling range or modulating range?
9. What is the periodic change in the controlled variable from one point in a process to another called?
10. What is the term that describes the delay in a control system?
11. What is the purpose of the final control element?
12. Name the six basic functions of a control system.
13. What are the two most common forms of energy used for control systems?

Chap. 2   Basic Control Theory   25

14. What component in a control system does the sensing and measuring?
15. Of what is the primary element usually made?
16. Why are electronic controls more sensitive than other types?
17. Why must relays be used with pneumatic controllers to obtain a definite on or off action?
18. Why is the timed two-position control preferred over the simple two-position control mode?
19. What causes the timed two-position control to operate?
20. In a timed two-position control, to what is the offset equal?
21. What is a sustained deviation between the point of the controlled variable which corresponds to the set point and the control point known as?
22. What types of controllers use bellows, diaphragms, or inverted bells immersed in oil?
23. What type of primary element is usually made from human hair, leather, or wood?
24. What device in the control system transforms the measurement of the controlled condition into a usable pulse?
25. What is a controlled motor, relay, or solenoid which converts the electric, electronic, or pneumatic energy into rotary, linear, or a switching action known as?

3

Basic Pneumatic Control System

Pneumatic control systems make use of compressed air to supply the required energy to operate the valves, motors, relays, and other required pneumatic control equipment. Therefore, the pneumatic control circuits are made up of air lines. Pneumatic control systems are made up of the following types of elements:

1. A reliable source of clean, dry compressed air. The air is generally stored in a receiving tank at some pressure capable of supplying all of the pneumatic devices which make up the system with enough air to operate properly.
2. A pressure reducing station which reduces the air pressure from the receiving tank to a normal operating pressure of 15–25 psig. The pressure depends on the system requirements.
3. Air lines, which may either be made from copper or polyethylene tubing, are used to connect the air supply to the controlling devices. These air lines are called the main lines.
4. The controlling instruments such as the thermostats, humidistats, and pressure-type controllers which are used to position the controlled devices.
5. The intermediate devices such as relays and switches.
6. All of the air lines which lead from the controlling devices to the controlled devices. These lines are called the branch lines.
7. The controlled devices, such as valves or damper operators. These devices may be called either operators or actuators.

The elements of a simple, single-pressure pneumatic control system are shown in Figure 3–1.

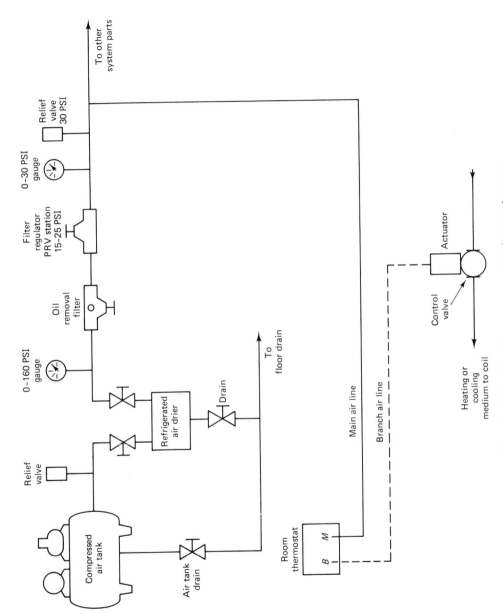

Figure 3-1 Typical single-pressure pneumatic control system.

The air compressor used in this system is an oil-free electrically driven unit. It has sufficient size so that it does not need to operate too frequently. The size is determined by the air requirements of the system. These compressors are normally designed so that they will operate not more than one-third of the time. Most of the pneumatic devices use air for their source of energy, and therefore they do, at times, bleed air into the atmosphere. The amount of air used by each one of the individual devices is added together to determine the air required for the system.

The speed of the compressor to be selected is also an important factor because of the possibility that the compressor may heat up and because of oil carry-over problems that could be encountered when the compressor operates at the higher speeds. This oil circulation can be detrimental to the controls unless extra oil separators are installed downstream of the compressor unit. The available electrical power must also be considered when making this choice. This electrical knowledge is also valuable when servicing starters, contractors, and other electrical equipment which is used in the system. The following formulas will apply when selecting an air compressor which will meet the requirements of the system:

CFM = total air consumption ÷ desired compressor operation (% running time) × 1728

Where

CFM = free cubic feet per minute of compressor capacity that will be required

Total air consumption = sum of the control devices used multiplied by the CFM for each of the respective devices

Compressor operation = percentage of the operating time required, such as $33\frac{1}{3}$ or 50% operation

1728 = the number of cubic inches per cubic foot

The need for clean, dry, oil-free air cannot be overemphasized to make sure that the air lines, controllers, switches, relays, restrictors, and other components in the system are kept clean and will operate in the manner for which they were designed. It is for this reason that other devices which will help to keep the air clean and dry will be included in the system.

The receiver tank should be equipped with a tank drain, and a refrigerated air dryer should be placed downstream of the compressor unit. A manual bypass valve for the refrigerated air dryer is not recommended because of the possibility that it could be opened and allow some, if not all, of the air to bypass the drier.

An oil removal filter is then placed downstream of the refrigerated drier to further ensure the delivery of oil-free air to the system. Next, the air flows to the pressure reducing device, and the pressure is reduced to the system operating pressure of 15–25 psi. A relief valve is placed in the air line next and set to open at 30 psi. This (30 psi) is the maximum safe operating pressure of most pneumatic devices, and this relief valve is meant to protect the system devices from damage.

After leaving the pressure reducing point, the air flows through the air lines to the controller. See Figure 3-1. In this figure the controller is the thermostat. The thermostat senses a change in the room air temperature and regulates its output pressure from 3 to 15 psi. This air signal is then transmitted to a control device, such as the valve actuator, through the branch line. The valve actuator receives this air pressure signal from the controller and causes the valve to be positioned between the extreme positions which are possible on the valve as the branch line varies from 3 to 15 psi.

This 3-15 psi variation in the controller output signal will normally occur within the throttling range of the thermostat (see Figure 3-1). If the throttling range of this thermostat is set for 4°F and the temperature set point is 72°F, when the ambient temperature is sensed by the thermostat as being 72°F, the branch line pressure is normally calibrated to 9 psi. The branch line pressure will be 3 psi when the ambient temperature is 70°F, and it will be 15 psi when the ambient temperature is 74°F.

ADVANTAGES OF PNEUMATIC CONTROL SYSTEMS

In most cases, pneumatic control systems are used to control heating and cooling equipment in commercial and industrial applications. Pneumatic control systems offer a number of very distinct advantages when used in these applications:

1. Pneumatic control equipment is inherently adaptable to a modulating type of operation; however, on and off operation can be realized with their use.
2. A large variety of control sequences and combinations can be realized by using relatively simple equipment.
3. Pneumatic control systems are normally quite free from operating problems.
4. They are very suitable where an explosive hazard exists.
5. The cost is usually less in large installations, especially where codes require that the low-voltage electrical wiring be installed inside of metal conduit.

REVIEW QUESTIONS

1. What type of air compressors are used in pneumatic control systems?
2. What is the recommended operating time for an air compressor for a pneumatic control system?
3. At what speed will the compressor for a pneumatic control system be more apt to pump the most oil?

4. Write the formula used for calculating the size of compressor needed for a pneumatic control system.
5. Why is clean, dry, oil-free air required for pneumatic control systems?
6. Is a manual bypass valve for the refrigerated air dryer recommended?
7. What is the normal operating pressure range of a thermostat?
8. Of what type of material are the air lines for pneumatic control systems made?
9. What are the air lines called that supply the air to the controllers in a pneumatic system?
10. What are the air lines called that lead from the controllers in a pneumatic system?
11. For what is the compressed air used in a pneumatic control system?
12. What types of elements make up pneumatic control systems?
13. What devices are used to prevent oil circulation through a pneumatic control system?
14. At what pressure does the air leave the pressure reducing station?
15. What is the maximum safe operating pressure of most pneumatic devices?
16. What is the output air pressure of a pneumatic room thermostat?
17. At what location is the oil removal filter in the pneumatic control system?
18. At what pressure is the relief valve in a pneumatic control system set to open?
19. To what type of operation is pneumatic control equipment inherently adaptable?
20. Between what points is the branch air line connected?

4

Pneumatic Control System Components

It is the purpose of any control system to provide a safe and stable operation of a process or a factory by maintaining the desired values of the variable condition under question or to provide the automatic control of a comfort air conditioning system. In reality, an automatic control system is a collection of the desired components, each having a definite function, and each is designed to interact with the other components in the control system, organized in such a manner that the system regulates itself.

PNEUMATIC CONTROLLERS

By nature, pneumatic controllers are inherently proportional in operation. Their design function is to regulate the flow of control system air in response to the changes which occur in the condition being controlled and to send an air signal to the controlled device—such as a valve or a damper actuator. Controllers are designed for either *direct acting* or *reverse acting*. They may be a thermostat, remote bulb or rigid stem temperature controller, a humidistat, or a pressure controller.

Direct-Acting Controller

A direct-acting controller is one which increases the branch line air pressure as it senses an increase in the temperature or pressure being controlled.

Reverse-Acting Controller

A reverse-acting controller is one which decreases the branch line air pressure as it senses an increase in the temperature or pressure being controlled.

The room thermostat is the most common type of pneumatic controller and is used more often than any other type of pneumatic control device. There are several different types of pneumatic thermostats, such as single pressure, dual pressure for summer/winter operation, and dual pressure used for day/night temperature control.

SINGLE-PRESSURE THERMOSTAT

A single-pressure thermostat is one which operates from a constant main line air pressure of 15–25 psi. A simple application of a single-air-pressure thermostat would be a thermostat which is used to control an actuator to regulate the heating, cooling, or ventilating functions of an air conditioning system. See Figure 4-1. There is, however, an exception to this condition, such as when a single thermostat is used to control more than one actuator at one time to provide a sequenced operation of the system. This use will be discussed later.

In operation, the main line air pressure enters the thermostat, passes through a filter, and then proceeds through a restrictor. A leak port is located just below the bimetal sensing element. The purpose of the leak port is to bleed air from the thermostat to the atmosphere to reduce its output pressure. The leak port is opened and closed by the action of the bimetal as it senses any changes in the temperature of the space. As the bimetal acts to close the leak port, the branch line air pressure increases. The main air valve, located inside the thermostat, responds to the opening and closing of the leak port through a mechanical lever mechanism, also in the thermostat, to allow the air to flow into the branch line or to close it off.

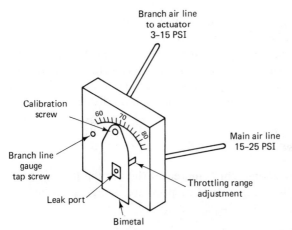

Figure 4-1 Single-pressure pneumatic thermostat diagram.

Chap. 4   Pneumatic Control System Components   33

Most of the pneumatic thermostats in use today provide a branch line gauge tap so that the thermostat output pressure may be read without removing it from the wall. See Figure 4-1. The sizes of the gauge tap fittings vary from manufacturer to manufacturer so it is important to have the proper calibration tools at hand.

To calibrate most single-pressure pneumatic thermostats, the following steps can be used:

1. Check to make certain that the main line is connected to the thermostat and that the branch line is connected to the actuator.
2. Check the main line air pressure to see that it is between 15 and 25 psi.
3. Connect the proper gauge port tap fitting to the gauge port on the thermostat.
4. Use an accurate thermometer and measure the ambient temperature at the thermostat. Take care not to touch the thermostat bimetal or get your hands too close to the thermostat so that the unit will not give a false reading.
5. Change the set point indicator to the ambient temperature just measured by using the special calibration wrench which is supplied by the thermostat manufacturer.
6. Turn the adjusting screw until the branch line air pressure reads 9 psi, or the midpoint of the spring range for the actuator which is being controlled.

The thermostat should now be in calibration. The temperature dial can now be set to the desired control point of the installation.

In operation, the controller governs the operation of the pnematic valve. See Figure 4-2. The controller can be of either the direct-acting or the reverse-acting type, and the valve can be either normally open or normally closed.

Let's suppose that the controller is direct acting and that the valve is normally open. See Figure 4-3. This combination of controller and valve is used for the flow control of some type of heating medium (either hot water or steam). Should there be a power failure, the valve would assume the full open position.

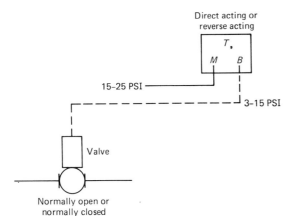

Figure 4-2 Basic temperature controller application.

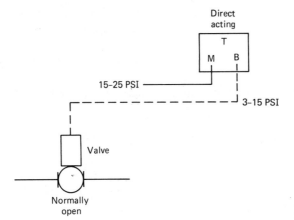

Figure 4-3 Normal heating arrangement.

When colder climates are encountered, this operation allows the heating medium to flow through the heating coils to prevent freeze-up in extremely cold weather and to provide heat to the space being conditioned.

If the controller in this example were reverse acting and the valve was the normally open type, this particular combination could be used for the control of a cooling medium. As the temperature sensed by the reverse-acting thermostat increases, its output air pressure decreases and allows the valve to open and the cooling medium to flow through.

If we compared an electric control system with the pneumatic system just described, the controller could just as well be an electric thermostat turning an electric element on and off. Basically, the pneumatic thermostat accomplishes the same thing with the exception that the output air pressure is varied to open and close the valve. In turn, the valve controls the flow of the heating or cooling medium through the coil in question.

Let's consider the combination of a controller and a damper actuator for a pneumatic control system. See Figure 4-4. In this type of application the controller sends an air signal to the damper actuator, which in turn positions the

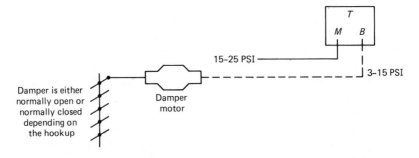

Figure 4-4 Basic damper control.

damper to properly control the flow of air through a coil, through a duct system, or into the conditioned space. In this application, normally closed or normally open refers to the position of the damper rather than to the damper actuator. If the damper had been located in the outside air stream, a normally closed type would have been used so that it would close in the advent of a power failure.

The use of a single-pressure thermostat controlling two valves in combination is often used to control both the heating and cooling in a single-air-pressure type of pneumatic control system. To obtain the proper sequence of operation, it is necessary to know the air pressure at which each of the valves is fully open and fully closed. Also, the valves must be identified as to whether they are normally open or normally closed. The components can be designated as follows: The thermostat is direct acting, valve 1 is normally open, and valve 2 is normally closed. See Figure 4-5.

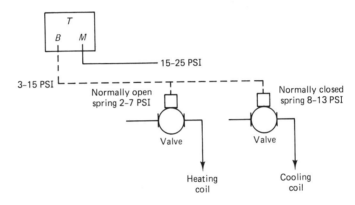

Figure 4-5 Valves in sequence for a heating and cooling system.

The spring ranges of the valves are also very important so that both valves cannot be open or closed at the same time. The spring ranges in our illustration have been split to eliminate this possibility. Valve 1 has a spring range of 2-7 psi. Valve 2 has a spring range of 8-13 psi. Thus, when the output pressure of the direct-acting thermostat is between 2 and 7 psi, valve 1 is closing and valve 2 is fully closed. When the output pressure of the thermostat reaches 8-13 psi, valve 2 is opening. When the output pressure is above 13 psi, valve 1 is fully closed and valve 2 is fully open. There is a dead spot between 7 and 8 psi which is used to prevent one valve from opening and the other from closing at the same time. This action eliminates the possibility of simultaneous heating and cooling.

In this example of sequencing valves, dampers could just as well have been used for one or both of the valves to control the flow of air into a space. However, the terms normally open and normally closed would apply to the position of the dampers rather than to the position of the actuators.

BASIC DUAL-PRESSURE PNEUMATIC SYSTEM

When dealing with dual-pressure systems, there are two different types that must be considered. The most common of the two is the summer/winter type of system, and the other is the day/night type of control system. Both of these types of control systems use different main line air pressures. The single-pressure system and the double-pressure system are the same from the compressor up to the pressure reducing station. Because there are two different main line air pressures used, the PRV station consists of two different pressure reducing regulators each having its own setting, resulting in different system operating pressures.

Summer/Winter System

The summer/winter type of system is one which provides for the seasonal demands for either heating or cooling. The medium required for the heating and/or cooling is supplied to the system by one supply line in combination with a return line. There may be either hot water or chilled water flowing through the system, depending on the seasonal requirements. Because the valve which controls this flow of liquid remains the same (either normally open or normally closed, but never both) through both seasons, the system must be equipped with a thermostat of either the direct-acting or the reverse-acting type. This function in the thermostat (direct acting or reverse acting) is done by changing the main line air pressure depending on whether heating or cooling is needed in the building.

In a typical dual-pressure type of control system, one pressure regulator will reduce the tank pressure to between 13 and 16 psi, and the other regulator will reduce the pressure to between 18 and 25 psi, depending on the control manufacturer's recommendations. The lower pressure is commonly used during the cooling season. The higher pressure is used during the heating season. In these applications it is necessary to use a summer/winter thermostat because there are two different air pressures that could be supplied to the thermostat, but only one at a time is used. There are two different bimetals in these thermostats; one is for the direct-acting section, and the other is for the reverse-acting section. When the lower pressure is supplied to the thermostats, the reverse-acting element has control over the output pressure of the thermostat. Also, when the higher pressure is supplied to the thermostat, the direct-acting bimetal has control over the output pressure.

For example, the two main air line signals from the PRV station are supplied to a three-way valve before entering the thermostat. See Figure 4-6. Also, either a manual or an automatic summer/winter switch is used. The purpose of this switch is to supply air pressure to the three-way air valve actuator to cause the normally closed port to open for heating and the normally open port to close for cooling. The air pressure on the valve actuator moves the valve to allow the lower operating pressure to flow out the common port and to the summer/winter thermostat. As the summer/winter switch is moved to the heating position, the

Chap. 4   Pneumatic Control System Components   37

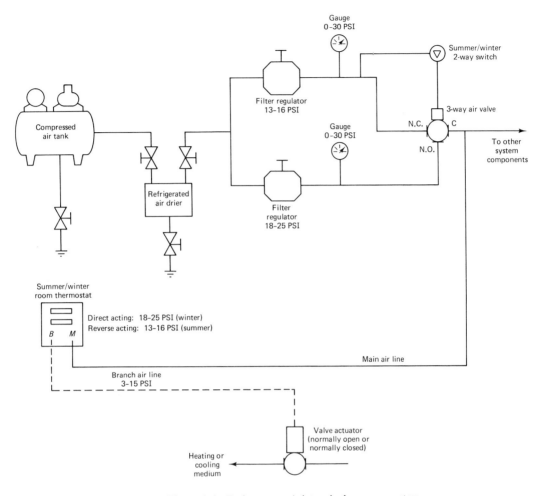

Figure 4-6 Basic summer/winter dual-pressure system.

air pressure is reduced on the valve actuator, causing the normally closed port to close and the normally open port to open, permitting the higher operating air pressure to flow to the thermostat.

The switching action of the summer/winter thermostat is done by changing the main line air pressure which is supplied to the thermostat. The switching action is an automatic, internal function of the thermostat. No manual adjustments are made at the thermostat to change it from direct acting to reverse acting. The only adjustment at the thermostat is the changing of the summer/winter switch, which changes the position of the three-way air valve, which changes the main line air pressure being supplied to the thermostat. This change can be accomplished by the use of a transmitter and a diverting relay. See Figure 4-7.

Because the summer/winter thermostat responds to two different main line

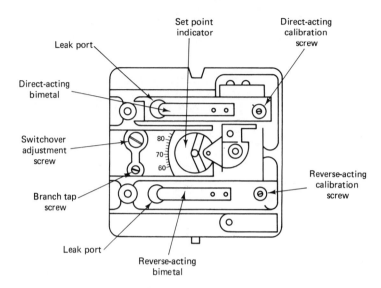

Figure 4-7 Typical summer/winter thermostat.

air pressures, the thermostat must be provided with some means of mechanically changing itself from direct acting to reverse acting with a change in the main line air pressure. This *switchover* is a factory adjustment which causes the thermostat to change as required. See Figure 4-7. When making a thermostat changeout, it is usually desirable to make this adjustment at the shop before going to the job site. To make this adjustment, simply connect an air line to the main thermostat port and supply a main air pressure to the thermostat equal to the switchover pressure required for the particular thermostat being replaced. Using the proper gauge tap adapter, set the temperature adjustment to the lowest setting. The reading on the gauge which is installed on the gauge tap should be approximately 15 psi. The switchover adjustment screw should then be turned counterclockwise until the branch line output air pressure drops to 0 psi. The switchover point is now set and is equal to the main line air pressure which is supplied to the thermostat. Do not confuse the main line air pressure which is supplied to the thermostat during the switchover adjustment with the main line air pressure that will be supplied to the thermostat while it is installed on the job. All this adjustment has done is to change the switchover pressure required by the thermostat so that it will change from summer to winter operation. Therefore, it is necessary to know the switchover pressure of the thermostat which is being replaced.

The calibration of dual-pressure thermostats is more complicated than the calibration of single-pressure units, because there are two bimetals and two main line air pressures being used. If the heating and cooling set points are to be the same, the branch line output air pressure should be approximately 9 psi at the set point regardless of which cycle the system is operating in. However, because the calibration procedure for summer/winter thermostats will be different for each manufacturer's unit, it is usually best to follow that particular manufacturer's recommendations for calibrating the thermostat.

Day/Night System

The day/night control system is designed for use on systems which require separate control points because of the varying occupancy and seasonal loads. This type of control system is especially adapted to school and hospital buildings. These types of control systems are used to control the temperature at various set points required for the day and night operation of the equipment. The day/night thermostat used on these systems is essentially the same as the summer/winter type of thermostat. See Figure 4-7. The main difference between the two thermostats is the two bimetals. In the day/night thermostat the bimetals are both direct acting, and in the summer/winter thermostat one bimetal is direct acting and the other is reverse acting. Each of the bimetals in the day/night thermostat is calibrated for different set points. When the main line air pressure is 13-16 psi (this may vary with different manufacturers), the day bimetal controls the output air pressure of the thermostat. When 18-25 psi air pressure is supplied (this may vary with different manufacturers), the night bimetal controls the output air pressure from the thermostat. As an example, the building may require a 75°F space temperature during the daytime hours and a nighttime temperature of only 65°F when the building is unoccupied. In this case the main line air pressure is changed in a similar manner to the summer/winter type of control system. This action allows the bimetal which is set to 65°F to take control of the system and control the temperature at the desired point. The switchover from one bimetal to the other is accomplished in the same manner as the action of the summer/winter thermostat.

Some of the day/night thermostats are equipped with a lever or button which allows each individual thermostat to be changed for daytime operation, while the remainder of the thermostats operate on the night cycle, as required by the central control switchover station. The thermostat may be returned to the nighttime operation by returning the switch or button to its original position. In some installations the thermostat will automatically return to the daytime operation when the control system is returned to day main line air pressure. The majority of the type of thermostats which have a built-in change-over control can be changed only to the control point of the daytime requirements and will return to the nighttime operating setting when the control system is on the nighttime operating pressure. Their control point cannot be changed to the nighttime setting when the main line air pressure requires daytime operation.

In operation, during the daytime operation 13-16 psi air pressure (this may vary with different manufacturers) is supplied to thermostat T_1. See Figure 4-8. In this example, a pneumatic-electric switch energizes both the fan, for continuous operation, and the electric-pneumatic relay. This allows the damper actuator,

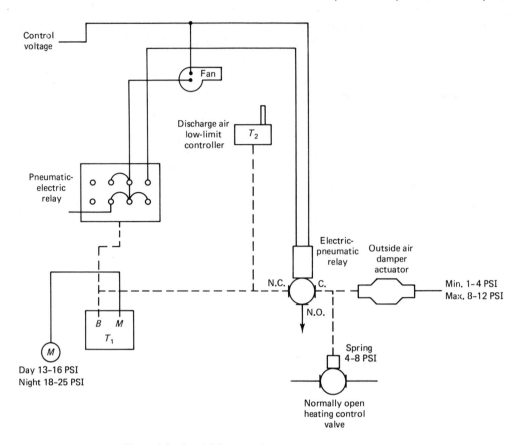

Figure 4-8 Day/night control unit ventilator cycle II.

along with the control valve, to be controlled by the day/night thermostat T_1. Should the temperature of the space drop below the thermostat set point T_1, the outside air damper will go to the fully closed position, and the control valve will go fully open to the coil. At this point, the coil is delivering the maximum amount of heating. As the temperature of the space approaches the set point of the thermostat T_1, the damper actuator moves the outside air damper to the nearly closed position. If the space temperature continues to rise, the control valve is at first modulated to the fully closed position, and then, if required, the damper actuator moves the outdoor air damper to the fully open position. This action allows the outdoor air to maintain the desired space temperature.

If the discharge air temperature should fall below the set point of the low-limit controller T_2, this controller will take control of the damper actuator and the control valve to maintain the minimum required discharge air temperature. When the thermostat is changed to the nighttime operation and the 18–25 psi air pressure (depending on the manufacturer's requirements) is supplied to the thermostat, the pneumatic-electric switch de-energizes both the unit fan and the electric-pneumatic relay. The outdoor air damper now returns to its fully closed position. The hot water valve, if used, will go to the fully open position. The thermostat will respond to a fall in the nighttime temperature and index the pneumatic-electric switch and energize the unit fan motor. The outdoor air damper will remain fully closed at this point. When the temperature of the space rises above the nighttime set point of thermostat T_1, the pneumatic-electric switch will de-energize the fan motor.

In our example, there was a special unit ventilator damper actuator used to position the outdoor air damper. The damper actuator for the unit ventilator is equipped with an internal spring arrangement which allows the actuator to operate gradually to allow a preset percentage of the piston stroke, hesitate at this point through a present pressure range, and then complete its full travel of the piston. The stroke of the actuator, before and after hesitation, is adjustable in the field application.

The split range of the unit ventilator damper actuator is usually 1-4 and 8-12 psi. In the previous illustration, as the signal pressure to the actuator is increased from 0 to 4 psi, the actuator shaft moves to a predetermined minimum position which is determined by the setting of the hesitation setting adjustment of the actuator. The actuator shaft remains at this position until the pressure has increased to 8 psi, and then upon the increase in pressure the actuator shaft is moved to the fully extended position.

TEMPERATURE CONTROLLERS

The temperature controllers which are used on pneumatic control systems are designed to perform many different control functions. Controllers of this type are usually of the remote bulb type or the rigid stem type. Some of their applications are the following: sensing outdoor air temperature, sensing return air temperature, sensing mixed air temperature, sensing water temperature, sensing discharge air temperature, and a multitude of other temperature control sensing functions. It is important to remember that regardless of the function of the controller it operates from the same operating pressures as the pneumatic-type thermostat discussed earlier and that the temperature controller is just a simple air pressure regulator, the same as the thermostat that has an air output signal pressure of 3-15 psi within the throttling range of the controller device.

Pneumatic temperature controllers can be used in many different ways, which leads to the confusion which surrounds pneumatic control systems. Temperature controllers differ from the basic room thermostat which is normally used only to sense any changes in room temperature and signal the equipment. The basic function is the most important fact to learn about a temperature controller. Regardless of how a controller is being applied, its function is to position some type of pneumatic actuator.

The basic temperature controller can be used as a limit control. See Figure 4-9. The limit control is used, in this example, to sense the discharge air

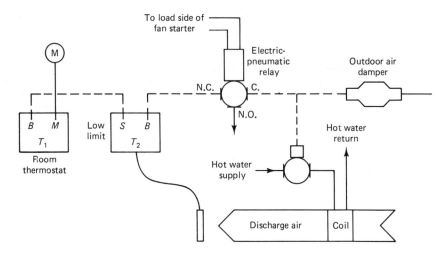

Figure 4-9 Temperature controller used as a low-limit controller.

temperature of the air handler and position the damper actuator and the control valve in case the discharge air temperature should fall below the minimum setting. If the discharge air temperature remains above the minimum setting, the thermostat T_1 positions the damper actuator and the control valve through the low-limit control. When the temperature of the discharge air remains above the set point of the low-limit control, then the air signal from the thermostat T_1 will pass through the limit control to position the actuators. Should the discharge air temperature fall below the minimum setting of the low-limit control, the action of the sensing element of the controller takes control of and will vary the output pressure. As the discharge air temperature rises above the set point of the controller T_2, the controller T_1 takes over the system control functions.

A controller with a remote bulb could also be used as a limit control. See Figure 4-10. A controller of the remote bulb type can usually be adjusted in the field to provide either direct- or reverse-acting functions. Basically the pneumatic controller is the same as the pneumatic thermostat. The difference is that the controller senses the air temperature with a remote bulb, while the thermostat senses the air temperature with a bimetal sensing element.

The set point and the throttling range can be adjusted on the temperature controller in Figure 4-10. Any adjustments on these controllers should always be determined by the application of the controller.

The unit temperature controller is another popular type of temperature controller. The use of these devices is usually limited to sensing and controlling the return air temperature in induction units, fan coil units, and unit ventilators. These types of controllers are available in direct-acting, reverse-acting, and reverse- or direct-acting types.

The difference between this type of controller and a room thermostat is that it is equipped with a remote bulb sensing element which is usually mounted in

Chap. 4 Pneumatic Control System Components 43

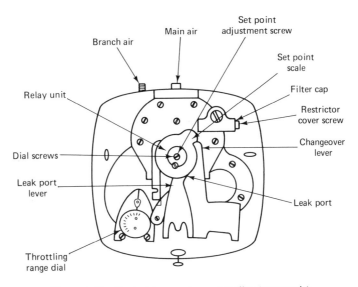

Figure 4-10 Typical temperature controller (pneumatic).

the return air duct to sense the return air temperature. These units can be used for both direct- and reverse-acting functions, the same as a summer/winter thermostat. These types of controllers have the ability to be either direct acting or reverse acting, depending on the main line air pressure which is supplied to the controller device.

This controller may be of either the direct-acting or the reverse-acting type. See Figure 4-11. This feature is determined by the main line air pressure supplied to the controller. In this example it is used to sense the return air temperature and position the three-way valve to control the flow through the coil to provide either heating or cooling. When heating is required, the controller will normally be in the winter cycle. As the temperature of the return air drops below the set point of the controller, the valve opens and allows the heating medium to flow through the coil and raise the temperature inside the space. When cooling is required, the controller is normally in the summer cycle. As the return air temperature

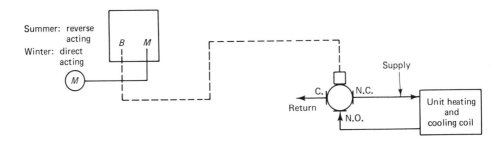

Figure 4-11 Dual-pressure unit temperature controller.

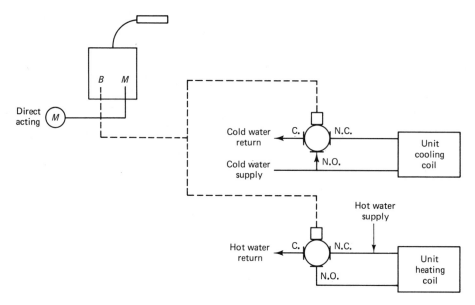

Figure 4-12 Single-pressure unit temperature controller.

rises above the set point of the controller, the three-way valve will open, allowing the cooling medium to flow through the unit coil and lower the temperature of the return air.

An important point about unit controllers is that when the controller is being used to control the flow through a single coil used for both heating and cooling, the controller should normally be capable of both direct action and reverse action. In an installation where separate coils are used for the heating and cooling process and each coil is equipped with its own control valve, the unit controller may be either a direct-acting-only or a reverse-acting-only valve. Then the action of the valves will depend on how the unit valves are piped to the coils.

The standard piping arrangement for separate coils is shown in Figure 4-12. In this particular case a direct-acting controller is used. The cooling control valve is normally closed to the coil, while the heating control valve is normally open to the coil.

PRESSURE CONTROLLERS

There are many types of pressure controllers which are available that will provide a great variety of pressure control functions. Two of the more common pressure controls are differential pressure control and constant volume control.

A simple application which provides differential pressure control uses a differential pressure transmitter as a static pressure sensing device. See Figure 4-13. The differential pressure transmitter is used as shown to sense and transmit a

Chap. 4   Pneumatic Control System Components   45

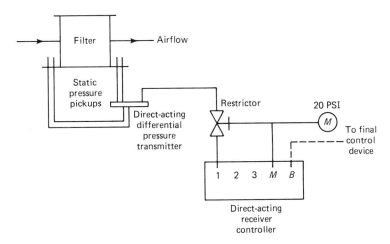

Figure 4-13 Typical differential pressure control.

differential pressure across air filters. It can just as well be used to sense the pressure drop across fans or coils, or between any two desired reference points. The differential pressure transmitter can be used, as illustrated, in conjunction with a receiver controller in the same manner as the room transmitter.

Another common control device is the volume control regulator. Its most popular use is for the control of the total airflow from mixing units which are used in high-pressure, high-velocity dual-duct air distribution systems. They can, however, also be used as velocity or static pressure controllers. See Figure 4-14. In this example, a volume control is being used as a constant volume controller. The cold actuator is operated by the volume controller, while the hot actuator is operated by the higher of the two signals from the room thermostat or from the volume controller. As the thermostat requires heat, the hot actuator is opened,

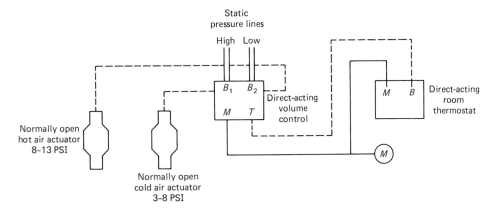

Figure 4-14 Typical constant volume control.

and the volume control closes the cold actuator. When the thermostat is satisfied, it will cause the hot actuator to be closed. The volume then detects the reduced airflow and gradually causes the cold actuator to open and maintain the required total airflow.

When the room thermostat requires full heating and the cold actuator is fully closed and the volume of total airflow becomes too high because of the excessive amount of hot duct pressure, the volume controller will take charge of the hot actuator and maintain a constant air volume. The hot actuator control is returned to the thermostat when its branch line air pressure becomes greater than that of the volume controller.

MASTER/SUBMASTER CONTROLLERS

A master controller is considered to be any type of pneumatic controller which transmits an output signal to another type of controller. The second controller, the submaster controller, is very similar to any other type of controller, the major difference being that the submaster's set point is changed upon any signal from the master controller; thus, it changes on every signal change from the master controller. Because of this feature, the usual application for the master/submaster controller system is best suited for some type of reset control. See Figure 4-15. This is a typical piping system consisting of a room thermostat which resets the set point of the submaster controller on either a fall or an increase in the room temperature.

This system is also suitable for a reset application for sensing a change in the unit discharge air temperature. See Figure 4-16. The branch signal from the room thermostat is piped directly to the reset port of the submaster being used as a temperature controller.

The resulting output signal from the submaster unit is piped directly to the final controlling device, which in turn controls the flow of the heating or cooling medium to the coil. The sensing element of the submaster controller is located in the discharge airstream of the unit. When used in this type of application, the discharge air temperature is varied in response to the room thermostat as the space temperature changes. Upon a change in the space temperature the signal

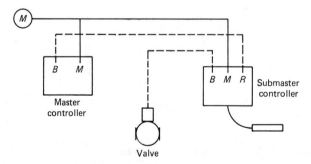

Figure 4-15 Basic master/submaster piping.

Chap. 4   Pneumatic Control System Components   47

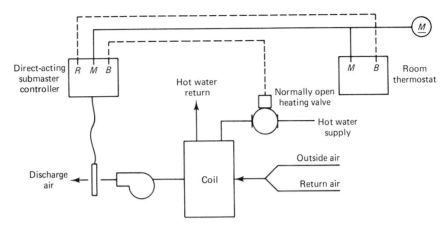

Figure 4-16 Typical reset control of unit discharge air temperature.

from the thermostat to the submaster also changes. These changes result in a lowering and raising of the set point of the submaster controller. The submaster in return senses the unit discharge air temperature and varies its output signal to the control valve to control the flow of the medium to the coil.

Master/submaster type of control is also used in applications requiring the resetting of the water supply temperature in response to the outside air temperature. See Figure 4-17. In this application, the master controller is a remote bulb type of controller with the sensing element in a location in order to sense the outside air temperature. The submaster controller is a rigid stem type of controller with the sensing element located in the hot water supply line to the unit.

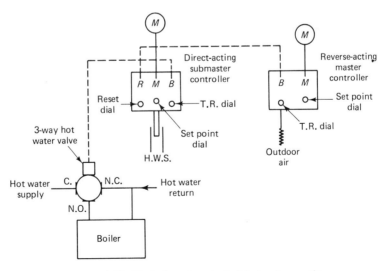

Figure 4-17 Typical reset control of hot water supply.

Typical reset schedule

O.D.	Output	H.W.S.
40°	15 PSI	210°
55°	9 PSI	160°
70°	3 PSI	110°

Figure 4-18 Typical reset schedule table.

On any application where a submaster controller is used, a reset schedule table should also be provided. See Figure 4-18. This is a typical reset schedule for the type of application shown in Figure 4-17.

The reset schedule indicates the water temperature which is to be maintained as the outside air temperature changes. Without the use of reset schedules, the controllers could not be properly calibrated.

In our example shown in Figure 4-17, when the outdoor air temperature drops to 40°F, the system requires that the temperature of the hot water be 210°F. The branch line air pressure from the master controller will be 15 psi when the outdoor air temperature reaches 40°F. The master controller in this application is set for reverse-acting operation. Therefore, as the controller detects a drop in the outdoor air temperature, its branch line air pressure increases. When the outdoor air temperature increases, the output signal from the master controller is decreased. This action results in a lowering of the set point of the submaster controller. Since the submaster is controlling a valve which is used for heating purposes and the valve is piped normally open to the boiler, the submaster is set for direct-acting operation. When the submaster sensing element detects a change in the temperature of the hot water supply, it varies the branch line air pressure to the heating valve. This in turn controls the flow of hot water from the boiler in response to the heating demands of the space.

The reset range required of the submaster controller determines where the submaster reset dial should be set. In our previous example, the required reset range is from 110 to 210°F. The span of this range is 100°F (210 − 110). The rest dial of this controller should be set at 100°F. The lowest temperature desired of the hot water is 110°F; this temperature becomes the set point for the submaster controller. Once the set point and the reset points have been adjusted as described, the master controller will now reset the submaster control from a minimum of 100°F up to 210°F and back down again to 110°F. The output signals from the master controller and the submaster control unit in this illustration are combined to reset upward from the established set point but never below it.

The reset chart shown in Figure 4-19 shows some of the other facts about this particular example. The entire range of the master controller (40-70°F) is representative of a throttling range of 30°F. The minimum output pressure (3 psi) and the maximum output pressure (15 psi) of the master controller occur within this 30°F throttling range.

In our example, a 10°F throttling range has been selected for the submaster controller. The output air pressure from the submaster control valve is 3-15 psi.

Chap. 4 Pneumatic Control System Components 49

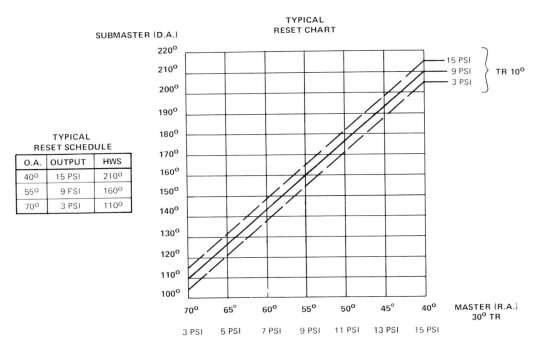

Figure 4-19 Typical reset chart. (*Courtesy of Johnson Controls, Inc., Control Products Division.*)

No matter what the temperature of the supply hot water, the minimum output air pressure of 3 psi and the maximum output air pressure of 15 psi of the submaster unit will occur within the 10°F throttling range. For example, when the outdoor air temperature drops to 55°F, the system requires 160°F hot water from the boiler. When the submaster has a 10°F throttling range, its output air pressure will be 3 psi at 155°F, 9 psi at 160°F, and 15 psi at 165°F. As the output air pressure from the master controller changes, all three lines shown on the reset chart can be read up and down the chart to determine the temperature of the hot water supply in a relationship to the outdoor air temperature.

RECEIVER CONTROLLERS AND TRANSMISSION SYSTEMS

In most modern pneumatic control systems the receiver controller is the controlling device. The sensing device which is used in combination with most receiver controllers in a transmission system is known as the transmitter.

In most of the older-style pneumatic control systems which were built around thermostats and master/submaster controllers the sensing elements and the control unit were installed together in a single unit. The old-style controllers had to be calibrated on an individual basis and adjusted at the point of installation and the point of control. The old-style controllers which used a remote bulb were effective when installed within the length of the capillary tube. Therefore, the controller could be mounted only as far away as the length of the capillary tube. When the transmission system is used, an air line is used between the receiver controller and the sensing device which may be as far away as several hundred feet. This flexibility permits the controller to be placed in a central control panel, and the sensing element can be located at some other point.

Another important difference between the older-style controllers and the newer transmission-type systems is that all of the adjustments for the transmission system are normally based on the air pressure, and on the older systems all of the settings were made based on temperature, humidity, etc. The relationship between pressure/temperature and pressure/humidity in the pneumatic control system should be understood.

The pneumatic controller, regardless of type, senses a change in the controlled condition (temperature or humidity) and changes the air signal in the connecting air line to the receiver controllers. The receiver controller then changes the branch line air pressure in response to the transmitter signal. The branch line is connected from the receiver controller to the controlled device just as in the old-style systems.

The receiver controller may be either of the direct-acting or the reverse-acting type. Some of the manufacturers use a changeover method at the receiver controller, and other manufacturers use an external reversing relay installed in the branch air line. Either of these methods provides the desired results by positioning the control device in the proper position.

Pneumatic-type transmitters are a one-pipe bleed-off type of device which makes use of a restrictor in the air supply line to help maintain the desired air pressure in the line between the transmitter and the receiver controller. A basic application of these devices makes use of a room transmitter and a receiver controller to position a heating valve. See Figure 4-20.

Most room transmitters have the same physical appearance as a standard room thermostat. The transmitter has only one air line connection between the transmitter and the receiver controller. It uses a bimetal element for sensing the room air temperature. When a change in temperature is sensed by the room transmitter, it will either bleed air to the atmosphere through its leak port, or it will close the leak port and cause an increase in air pressure in the transmission line. The transmitter works in combination with the external restrictor which has a 20-psi main line air pressure supply to maintain an output signal of 3-15 psi in the transmission line connected to the receiver controller. This 3-15 psi signal varies over the range of the transmitter. Standard room transmitters have a range of 50-90°F. In normal installations the transmitter is a direct-acting type, and at 50°F ambient temperature the transmission line air pressure would be 3 psi, and at the 90°F ambient temperature it would be 15 psi. The reciever controller will then vary its output signal based on the setting of the controller.

Transmitters, like any type of temperature controller, are available in a wide variety of temperature ranges. There will always be a direct relationship between

Chap. 4   Pneumatic Control System Components   51

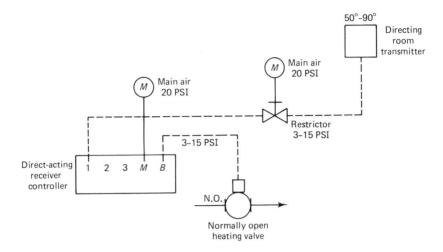

Figure 4-20   Room transmitter/receiver control.

the temperature span of the transmitter and the output span of the transmitter in psi; it will always be necessary that we know the temperature range of the transmitter so that the proper adjustments can be made on the receiver controller. Their relationship is the high end of the range minus the low end of the range. *Example:* A 0-100 °F transmitter has a span of 100 °F. The output span would be the 15 psi minus the 3-psi or 12-psi output span.

The sensitivity of the controller is the psi change in the transmitter output signal for each degree of change of the temperature being sensed. The sensitivity of the transmitter can be calculated as follows:

$$\text{transmitter sensitivity} = 12 \text{ psi/span}°$$

Example: If we assume a room transmitter has a 50-90 °F range (see Figure 4-20), the temperature span of this room transmitter is 40 °F. We then divide the temperature span of 40 °F into the output span of 12 psi; the sensitivity is 0.30 psi for each degree calculated. Thus, for each and every degree temperature change sensed by the room transmitter, the output signal from that transmitter will be varied by 0.30 psi of air pressure.

The sensitivity chart shown in Figure 4-21 indicates the precalculated sensitivities for most Robertshaw transmitters.

Because the throttling range, the proportional band, for pneumatic transmitters is set at the receiver controller, other information is required. The throttling range is the number of degrees that the controlled medium must change for the receiver controller to change the branch line air pressure between the 3-15 psi. This is the same requirement as that for a standard temperature controller. In most cases the throttling range of a controller is stated in degrees of temperature. The throttling range adjustment on a receiver controller is usually designated in either psig or as a percentage, making it necessary to convert the desired throt-

Transmitter	Range	Span	Sensitivity
HP2232	30%–80% RH	50%	0.24 psi/% RH
PP2323	0"–2" WC	2"	6.0 psi/" WC
PP2323	0"–7" WC	7"	1.7 psi/" WC
TP2220	50 to 90°F	40°	0.30 psi/°F
TP2252	40 to 140°F	100°	0.12 psi/°F
TP2252	0 to 100°F	100°	0.12 psi/°F
TP2252	40 to 240°F	200°	0.06 psi/°F
TP2252	−40 to 160°F	200°	0.06 psi/°F
TP2252	−25 to 125°F	150°	0.08 psi/°F
TP2254	40 to 100°F	60°	0.20 psi/°F

Figure 4-21 Sensitivity chart.

tling range from degrees to psi or percentage. Using the preceding example, a room transmitter having a 0.30-psi sensitivity; we assume a desired throttling range of 3°F. Multiply the 3°F by 0.30 psi to obtain a 0.9-psi throttling range. If the receiver controller adjustment range is indicated in psi, simply set the adjustment to 0.9 psi. In cases where the adjustment must be made as a percentage, use the chart shown in Figure 4-22. A percentage can be determined from this chart which closely equals the psi throttling range. As an example, an 0.9-psi throttling range would be equal to approximately a 5% throttling range. In this case, set the throttling range adjustment to 5%.

It should be remembered that the correct throttling range can only be determined by the particular requirements within each system. When too large a setting is used, a large deviation from the set point will be experienced during periods of load changes. Also, if the throttling range is set too close, the control system will experience hunting conditions. The throttling range should be adjusted on each installation to just prevent a hunting condition.

Receiver controllers can be used to accept either a single input signal (see Figure 4-23) or as a dual control which receives signals from two different transmitters (see Figure 4-24). Some units have other connections available which will allow a remote set point adjustment. Some control manufacturers make use of separate units for single and dual applications, while other manufacturers use a single unit which is capable of providing either function.

A typical example of a receiver controller with the throttling range adjustment shown in percentage form is shown in Figure 4-25. Most of these controllers also have a local control point adjustment and a control point scale which may either be in psi, temperature, or humidity. Also, there are a number of different

T.R. in PSI	0.5	1.2	1.87	2.5	3.75	5.0
T.R. in %	4	10	15	20	30	40

Figure 4-22 Transmitter receiver PSI and percentage chart.

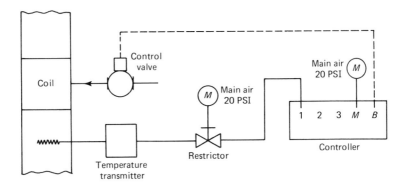

Figure 4-23 Typical single-input receiver controller.

connections to the receiver controller, as follows: The connection labeled 1 is normally used for the primary transmitter signal, the 3 port is for the secondary transmitter signal, the M port is for the main air connection, and B is for the branch line connection to the control device. Different manufacturers may use different designations for the ports on their units. The 2 port is used for the connection for a remote manual set point adjustment if it is desired.

When a receiver controller is being used in a dual-input application, there is another adjustment provided that determines the effectiveness of the secondary transmitter signal of the controller. This adjustment is normally called the authority adjustment. It may be better understood if it is called the reset adjustment. Thus, for every degree of temperature change in the outdoor air temperature the hot water supply temperature will be changed proportionally.

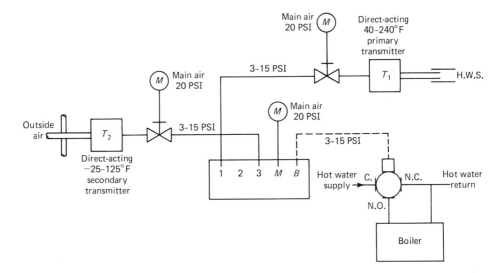

Figure 4-24 Typical dual-input receiver controller.

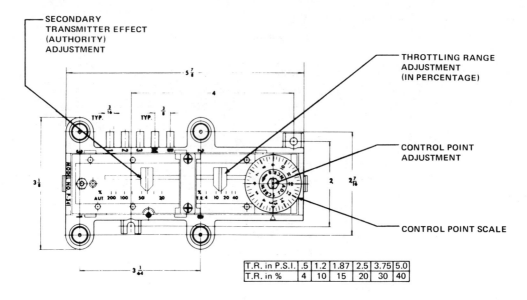

RECEIVER CONTROLLER

Figure 4-25 Receiver controller. (*Courtesy of Johnson Controls, Inc., Control Products Division.*)

A single-input receiver controller may be used for the control of the discharge air temperature and to position a control valve in response to the requirements. See Figure 4-23. Also, a typical master/submaster controller application may be for the control of the hot water supply temperature used with a receiver controller and two temperature transmitters (see Figure 4-24), thus providing the reset control of the hot water supply and accomplishing the same thing as the master/submaster reset control as shown in Figure 4-17 and using the old-style type of controllers. However, the more modern type of receiver controller transmitter control concept allows a more centralized control point, and closer control adjustments are possible.

When used for hot water supply temperature control reset, the primary transmitter senses the hot water supply temperature and transmits a signal to the 1 port of the controller. The secondary transmitter used to provide the required reset is piped to port 3 on this controller. There are restrictors which are placed in both of the transmission lines. Some control manufacturers use internal-type restrictors in the receiver controller. However, the use of external restrictors allows the transmitters to be placed at some distance from the controller. Also, it is easier to replace a clogged restrictor when it is installed externally.

A reset schedule must be used when setting up this type of system, the transmitters must be selected, and the throttling range and the reset percentage must be available. The first step in setting up the receiver controller is to select the proper and desired transmitters. In our example in Figure 4-17, an outdoor air transmitter with a −25–125 °F temperature range is selected, and a hot water supply transmitter with a 40–240 °F temperature range can be used to provide the desired results.

The second step is to determine a throttling range. Using past experience and the knowledge of the amount of pressure drop present when sizing the hot water control valve will allow the selection of a probable throttling range. In our example, a throttling range of 20 °F is used for this purpose. Based on using a 200 °F primary transmitter with a 200 °F span (240 − 40 °F), each degree of change in the hot water temperature will represent an 0.06-psi change in the transmitter output pressure signal. Thus, 20 °F × 0.06 psi provides a throttling range setting of 1.2 psi of air pressure or a 10% setting as shown in the conversion chart in Figure 4-22.

Step three is to determine the reset (or authority) percentage. This calculation may be made from information obtained from the available reset schedule. See Figure 4-26. This schedule shows what the setup pressures would be for this application. These pressures have been determined by using an output pressure chart for transmitters. See Figure 4-27. When it is desired to know the output psi at ports 1 and 3 for any given temperature, we must refer to the output pressure chart. The outdoor air transmitter is rated at −25–125 °F. As we refer to the output pressure chart for this transmitter at a 0 °F outdoor temperature, the transmitter will supply an air pressure of 5 psi to port 3. In this method, we can determine any pressure/temperature relationship for this particular transmitter from the chart. The hot water supply transmitter is rated at 40–240 °F. The output pressure chart for this transmitter indicates a pressure at port 1 to be 12.6 psi with a supply hot water temperature of 200 °F.

When all of this information has been determined, we can use the following formula to calculate the reset percentage:

$$\text{authority \%} = \frac{\text{change in pressure at port 1} + \text{T.R. psig}}{\text{change in pressure at port 3}} \times 100$$

When referring to the reset schedule, the change in pressure at port 1 (difference between the lowest and the highest psi or 12.6 psi − 6.6 psi) of the hot water supply transmitter is 6.0 psi. The change in pressure (difference between the highest and the lowest pressures to port 3 or 10.6 psi − 5.0 psi) is 5.6 psi. When we add the throttling range, which is 1.2 psi, this gives us 7.2 psi in the numerator of the formula. By dividing this 7.2 psi by 5.6 psi and multiplying the results

Example	Port 3 pressure	Outdoor air temp.	Hot water temp.	Port 1 pressure
Set point	5.0 PSI	0° F	200° F	12.6 PSI
	7.8 PSI	35° F	150° F	9.6 PSI
	10.6 PSI	70° F	100° F	6.6 PSI

Figure 4-26 Reset schedule.

	Temperature range (°F)					Output (PSI)
	−25−125°F	0−100°F	40−140°F	−40−160°F	40−240°F	
Actual temperature	−25	0	40	−40	40	3.0
	−22	2	42	−36	44	3.24
	−19	4	44	−32	48	3.48
	−16	6	46	−28	52	3.72
	−13	8	48	−24	56	3.96
	−10	10	50	−20	60	4.2
	−7	12	52	−16	64	4.44
	−4	14	54	−12	68	4.68
	−1	16	56	−8	72	4.92
	2	18	58	−4	76	5.16
	5	20	60	0	80	5.4
	8	22	62	4	84	5.64
	11	24	64	8	88	5.88
	14	26	66	12	92	6.12
	17	28	68	16	96	6.36
	20	30	70	20	100	6.6
	23	32	72	24	104	6.84
	26	34	74	28	108	7.08
	29	36	76	32	112	7.32
	32	38	78	36	116	7.56
	35	40	80	40	120	7.8
	38	42	82	44	124	8.04
	41	44	84	48	128	8.28
	44	46	86	52	132	8.52
	47	48	88	56	136	8.76
	50	50	90	60	140	9.0
	53	52	92	64	144	9.24
	56	54	94	68	148	9.48
	59	56	96	72	152	9.72
	62	58	98	76	156	9.96
	65	60	100	80	160	10.2
	68	62	102	84	164	10.44
	71	64	104	88	168	10.68
	74	66	106	92	172	10.92
	77	68	108	96	176	11.16
	80	70	110	100	180	11.4
	83	73	112	104	184	11.64
	86	74	114	108	188	11.88
	89	76	116	112	192	12.12
	92	78	118	116	196	12.36
	95	80	120	120	200	12.6
	98	82	122	124	204	12.84
	101	84	124	128	208	13.08
	104	86	126	132	212	13.32
	107	88	128	116	216	13.56
	110	90	130	140	220	13.8
	113	92	132	144	224	14.04
	116	94	134	148	228	14.28
	119	96	136	152	232	14.52
	122	98	138	156	236	14.76
	125	100	140	160	240	15.0

Figure 4-27 Output pressure chart.

by 100, the rate of the reset percentage will be 129%. Therefore, the authority adjustment on the receiver controller will be set at this percentage. In any case, the receiver controller must be calibrated, and this procedure as well as the determination of the reset percentage should be done in accordance with the particular control manufacturer's setup procedures.

In the example just discussed, the transmitter is used to sense static pressure in inches of water column (H_2O). When a differential pressure is sensed between the high-and the low-pressure sensing lines of the transmitter, the transmitter will transmit a 3–15 psi signal which is proportional to the pressure which is sensed at the receiver controller. The receiver controller will in turn position the final control device. In an application of this type, the tubing size of the sensing lines is very important. The sensing lines must be sized according to the length of line running to the transmitter.

Most of the differential pressure type of controllers are nonadjustable. Because of this, a control must be selected which will sense the static pressure that is within the desired limits of the application.

There are other types of differential pressure controllers which are available for use in monitoring applications that require higher air pressures, such as those maintaining a constant pressure difference across the supply and return main lines used in a forced circulation water-type system. Applications of this type normally are provided with two-way valves which control terminal units and the final control device, which is located either in a pump bypass or directly in the pump discharge line.

REVIEW QUESTIONS

1. What is the purpose of any control system?
2. What is the function of pneumatic controllers?
3. What type of controller increases the branch line air pressure on an increase in temperature or pressure?
4. What type of controller decreases the branch line air pressure on an increase in temperature or pressure?
5. In what type of installation are dual-pressure controllers used?
6. In a pneumatic thermostat, what does the air pass through when going from the line to the thermostat?
7. In a pneumatic thermostat, what does the bimetal do in response to a temperature change?
8. What is the advantage of a heating valve assuming the full open position in the case of a power failure?
9. Why is it desirable to use the proper spring ranges when installing pneumatic valves?
10. When discussing dampers, to what do the terms normally open and normally closed refer?
11. What is the difference between the summer/winter thermostat and the day/night thermostat?

12. In a dual-pressure pneumatic control system, during which season is the lower pressure commonly used?
13. What is the purpose of the summer/winter switch in a pneumatic control system?
14. To what do the letters PE refer?
15. What is the difference between the pneumatic thermostat and the pneumatic temperature controller?
16. What is the basic function of a pneumatic controller?
17. What is an important point about pneumatic controllers?
18. Name two of the more common types of pressure controls.
19. In what applications is the differential pressure transmitter used?
20. What is a master controller?
21. What is the difference between the submaster controller and any other type of pneumatic controller?
22. For what type of application is the master/submaster controller system best suited?
23. What determines where the submaster reset dial should be set?
24. What is the controlling device in most modern pneumatic control systems?
25. What flexibility permits the controller to be placed in a central control panel and the sensing element to be located at some other point?
26. What two relationships in a pneumatic control system should be understood?
27. What types of devices are pneumatic transmitters?
28. What is the first step in setting up a receiver controller for reset operation?
29. Write the formula for calculating reset percentage.
30. Are differential pressure types of controllers adjustable?

5

Pneumatic Control Valves

Pneumatic control valves are used as a means for controlling the flow of heating or cooling media. The flow which is being controlled may be steam, water, or some combination of these two media. It is essential that these devices be designed and sized so that the desired results can be obtained from an air conditioning system.

A pneumatic-type motorized valve is made up of the following components:

Actuator: This is the part of the automatic control valve which causes the valve stem to move.

Valve body: This is the part of the automatic control valve in which the medium is contained.

Disc: This is the movable part of the valve that makes contact with the valve seat when the valve is closed and which varies the orifice to control the flow. Discs are sometimes built so that the part of the disc that comes in contact with the seat may be replaced. This type of disc is called the *renewable disc*.

Guide: This is the part of the valve disc that keeps the disc aligned with the valve seat. Top or bottom guides or both are used to accomplish this centering function.

Port: This term makes reference to the flow controlling opening between the valve and its seat when the valve is wide open. It does not refer to the body size or the end connection size. Standard valve ports are the sizes normally used in the valve. Reduced ports have flow controlling areas equal to the valve port area of valves with the same body size as the reduced port size.

Trim: The trim consists of all the parts of a valve that are in contact with the flowing medium but are not part of the valve shell or casting. The seats, discs, throttling plugs, stems, packing rings, etc., are all trim components.

Pneumatic control valves have classifications according to their body design. These classifications are known as two-way valves, three-way valves, single seat, or double seat. When the control actions of the valve are considered, they are called normally open, normally closed, and three-way mixing diverting types. These positions refer to the position that the valve is in when there is no air pressure being applied to the operator. Thus, a normally closed valve remains closed when there is no air pressure applied to the operator. When three-way valves are being considered, these terms refer to the valve ports rather than to the position of the entire valve.

In most applications, the normally closed valves are installed on heating equipment including unit heaters, radiators, and heating coils. The normally closed type of valve is normally used on installations such as humidifiers, water chillers, or other types of cooling applications which require this type of valve. These types of valves are sometimes used in combination or in some sequence for sequencing operation.

Single-seated valves are equipped with only one seat and one disc. They are less expensive than the double-seated valves because of the extra expense in the manufacture of the double-seat arrangement. Single-seated valves are suitable where a tight shutoff of the line is required. See Figure 5-1. The single-seated valve will always require more force to close it than double-seated valves of the same size because there is no balancing force to help it to close against the fluid pressure against the valve disc.

The seats and discs in a double-seated valve are arranged in such a manner in the closed position that there is very little fluid pressure forcing the valve toward the open or the closed position. A double-seated valve will require less power

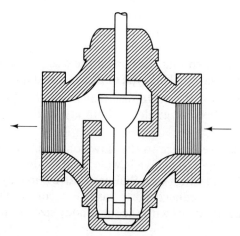

Figure 5-1 Normally open single-seated valve.

Chap. 5   Pneumatic Control Valves

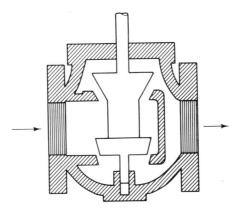

Figure 5-2   Normally open double-seated valve.

for operation than a single-seated valve of the same port area and size. See Figure 5-2. Another advantage that the double-seated valves have is that they will usually have a larger port area for any given pipe size. One disadvantage is that they do not always shut off tightly because both of the seats are rigidly connected together and the changes in temperature of the controlled fluid will cause the disc or the valve casting to expand, allowing one disc to seat before the other one, which often allows the fluid to leak past the valve.

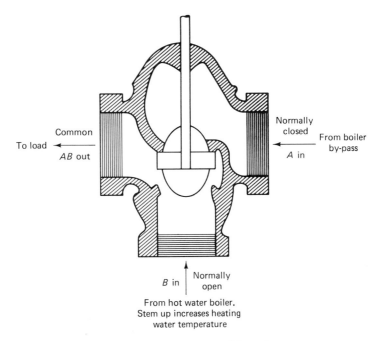

Figure 5-3   Three-way mixing valve.

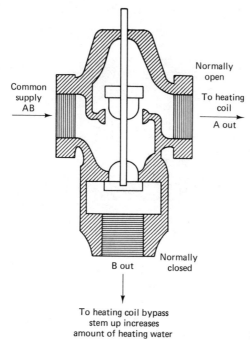

Figure 5-4 Three-way diverting valve.

Three-way mixing valves are equipped with three pipe openings. There will always be two inlets and one outlet pipe connection. The amount of fluid that enters each of the two inlets can be varied by moving the valve stem. See Figure 5-3. Valves which are designed for mixing purposes are also generally suitable for diverting applications.

Three-way diverting valves are always equipped with one inlet and two outlet pipe fittings. Thus, the fluid entering the inlet connection can be diverted to either of the two outlet connections in any desired proportion by moving the valve stem. See Figure 5-4.

VALVE FLOW CHARACTERISTICS

A valve flow characteristic is a relationship between the valve stem travel, expressed in percent of travel, and the flow of the controlled fluid through the valve, expressed in percent of full flow.

Quick Opening

This action of quick opening provides the maximum possible flow as soon as the valve starts to open. In most cases this type of valve is equipped with a flat seat and no throttling plug. These valves are most useful for two-position control.

Linear

Under linear conditions, regardless of the amount of valve opening, the valve opening and the flow are in direct relation.

Equal Percentage

In the equal percentage relationship, for each equal increment of valve opening, the flow is increased by an equal percentage. *Example:* Suppose we have a valve stem which lifts 30% of its total travel and allows a flow of 4.0 GPM. Then suppose that the valve stem moves an additional 10% of its travel so that it now is 40% open. The fluid flow will now be 6.4 GPM, for a 60% increase in the fluid flow. Should the valve stem move another 10% toward the open position, it would now be 50% open. The fluid flow will now be 10.2 GPM. This is another 60% increase in the fluid flow. Equal percentage discs are useful in control applications where from time to time there are wide variations in the load. Valves which are equipped with equal percentage flow characteristics are almost always used on hydronic systems.

VALVE RATINGS AND TERMINOLOGY

The following is a listing of the valve ratings and related terminology.

Capacity Index

The capacity index is the amount of water in gallons per minute at 60°F that will flow through a given valve with a pressure drop of 1 psi. This is also known as the flow coefficient of the valve. The symbol used for flow coefficient is C_v. When the C_v of the valve has been determined, the flow rate of any type of fluid through the same valve can be calculated, provided the characteristics of the fluid and the pressure drop through the valve are known.

Close Off

The valve close-off rating is the maximum allowable pressure drop that the valve can be subjected to while it is fully closed. This rating is a function of the power which is available from the valve actuator by holding the valve closed against this pressure drop. This rating is independent of the actual valve body rating.

Close-off Rating for Three-Way Valves

For three-way valves, the close-off rating is the maximum pressure difference between the outlet and either of the two inlets for mixing valves or the pressure difference between the inlet and either of the two outlets for diverting valves.

Maximum Fluid Pressure and Temperature

The maximum fluid pressure and the temperature ratings indicate the maximum pressure and temperature limitations which are placed upon a valve. They are the maximums to which the valve may be subjected. The rating is determined by the type of packing, body material, or disc material or the actuator limitations.

Pressure Drop

The pressure drop is defined as the difference between the fluid pressure entering the valve and the fluid pressure leaving the valve.

Critical Pressure Drop

The flow of fluid through a valve is increased with a resulting increase in pressure drop until a critical value is reached. Thus, any pressure drop in excess of the rated critical pressure drop must be avoided to make sure that noise-free operation and a minimum amount of valve wear is experienced.

Body Rating

The body rating is defined as the correlation between the safe permissible fluid flow pressure and the flowing medium temperature. Every nominal valve body rating has a definite and corresponding permissible pressure at the various temperatures that may be encountered. Therefore, the critical pressure is dependent on the fluid temperature.

Nominal Body Rating

The nominal body rating is the nominal pressure rating of the valve body and is expressed in psi. Usually this rating is cast into the side of the valve body. The purpose of this rating is to provide a convenient method of classifying the valve by pressure class for proper identification.

Rangeability

The rangeability of a valve is defined as the ratio of the maximum controllable fluid flow to the minimum controllable fluid flow. *Example:* Consider a valve with a rangeability of 50-1 and having a fluid flow capacity of 100 GPM when fully open. This valve can control a flow down to 2 GPM accurately. This type of valve may or may not have a tight shutoff seat. The changes in the flow of the fluid from 0 to 2 psi are accomplished with a very little amount of stem movement.

CLOSE-OFF RATINGS AND SPRING RANGES

The pressure drop which acts against the unbalanced area of a valve produces a thrust in this area. The actuator must overcome this thrust through the application of an additional pressure signal exerted above the top end of the signal range for normally open types of control valves or by reducing the signal pressure below the bottom end of the range for normally closed control valves. Normally open control valves which are usually used on heating applications usually use lower spring ranges. *Example:* When a 4-8 psi spring range valve is used, it is fully open when 0-4 psi are applied to the actuator. When 4-8 psi are applied, the valve begins to close. When the pressure reaches 8 psi, the valve is fully closed. When the available air pressure from the air reducing station is 15 psi, this allows 7 psi to hold the valve closed against the fluid flow. The span of the range spring is actually increased by the amount of pressure drop across the valve seat. If the pressure drop across the valve seat should go above the close-off rating of the valve, it may not close off properly.

Valves with higher spring ranges are normally used in conjunction with the normally closed types of valves. *Example:* A valve with an 8-13 psi spring range provides the additional pressure required for proper valve operation.

Not all manufacturers use the same spring ranges. For example, a particular manufacturer may design his actuator with a 3-7 psi spring range while another may use a 4-8 psi spring range. These differences are very minute and insignificant. The only exception to this insignificance would be when valves are being sequenced for heating and cooling applications. The heating valve may have a 3-7 psi spring range and the cooling valve may have an 8-13 psi spring range. The dead spot between 7 and 8 psi is built into the design to prevent one valve from opening and the other from closing at the same time, thus preventing the possibility of having cooling and heating in operation at the same time. Some manufacturers provide a means for making slight adjustments on applications when the spring ranges are too close or when they overlap. In some cases a proportioning relay can be used to aid the valve in moving exactly as far as the controller requires.

STEAM VALVES

Valves which are used for the control of steam should not offer any more resistance to the flow than any other element in the system so that the flow can be controlled effectively. Because of this, the valve selection for the proper pressure drop should be the first step in selecting the pressures which are to be used in the system.

Some of the things that should be considered when selecting a steam valve pressure drop are as follows:

The valve pressure drop in a modulating-type system should be at least 80% of the total difference between the supply and the return main line pressure. Some exceptions are the following:

1. The 80% drop exceeds 50% of the absolute upstream pressure. In this case the 50% of the absolute upstream pressure should be used as the valve pressure drop.
2. Due to some special circumstances, an 80% drop would result in a steam pressure in the heating unit that is too low. As an example, consider pressures which are to be selected for a new building that is to be equipped with standing radiation. The heating radiators are to be selected to produce the designed quantity of heat when they are filled with 1 psi of steam pressure, inside the radiator, to become 20% of the difference between the supply and the return main line pressure. If 5 lb is selected as the main line supply pressure, then 80% × 5 lb = 4 lb of valve pressure drop. This will allow an 80% pressure drop through the valve and still allow 1 lb of steam pressure to be in the radiator when the valve is completely open.

Remember that in most installations a loss of heat output required to use an 80% valve drop is quite small. As an example, a radiator which is supplied from a 5-psi boiler through a valve having a 1-psi (20%) pressure drop will produce 100,000 Btu/hr. The radiator would be operating with a 4-psi steam pressure. If a smaller valve were used on this same radiator, a valve with a 4-psi (80%) pressure drop would produce 92,000 Btu/hr, a loss of only 8%.

Due to this fact, it is quite possible to use a fairly low supply main line pressure and still use 80% of the supply to return main line pressure for valve pressure drop without jeopardizing the heat output of most steam-supplied devices.

Another example would be to use only a 2-lb pressure supply main line steam pressure with a 4-in. vacuum in the return lines. The difference between the supply and the return main pressures is 4 psi. Thus, 80% × 4 psi = 3.2 psi, the most favorable valve pressure drop. This leaves a pressure of approximately 2.5-in. vacuum. A steam coil which is completely full of steam at 2.5-in. vacuum will have a heat output of about 92% of the full output based on the 2 psi.

SUPPLY AND RETURN MAIN PRESSURES

As stated previously, the supply main line pressure should be high enough to allow an 80% pressure drop through the control valve and still leave enough steam pressure to produce the desired heat output of the unit. The steam supply main line pressure should be held as steady as possible. Should the boiler pressure not be sufficiently constant, a pressure reducing valve should be installed in the inlet lines of all the steam appliances in which the output temperatures will vary rapidly with any fluctuations in the steam supply pressure.

Steam return main line pressure should also remain constant if at all possible, because any variation in the return line pressure will cause a fluctuation in the flow of steam through the control valves, even though the valves may not change their positions. Thus, from the standpoint of control, atmospheric-type return lines which use a condensate pump are better systems than the vacuum return system which has a vacuum pump that can cycle over a range of several inches of vacuum.

DESIGN CONSIDERATIONS FOR STEAM COILS

The general rules for the most favorable design conditions and piping arrangement that provides the best control valve performance are as follows:

1. The main line pressure should be held at the design pressure plus or minus 1 psi.
2. The returns should be at atmospheric pressure unless lifts are required in the return lines.
3. The traps should be sized to pass the design flow of condensate at 1 psi of pressure drop.
4. An equalizer line should be used to prevent the formation of a vacuum inside the coil.
5. The control valve pressure drop should be 80% of the difference between the supply and the return main line pressures.

Fluctuating supply and return pressures can have a great effect on the operation of the system. As an example, assume that we have a system where the boiler cycles off at 6 lb and on at 2 lb of steam pressure. This same system has a vacuum pump that cuts in at 4-in. vacuum and cuts off at 8-in. vacuum. Therefore, the pressure difference between the supply and the return lines could possibly vary from a minimum of 4 psi to a maximum of 10 psi of pressure as the boiler and the vacuum pump go through their periodic cycles. This indicates a 59% capacity variation in the control valves as the system pressures vary. The control valves that were originally and correctly designed for a 4-psi pressure drop are 59% too large during the periods when there is a 10-lb pressure drop between the steam supply and the return lines.

An exception to this situation would be a high-vacuum system because its purpose is to lower the steam temperature and pressure as the heating load is decreased. Vacuum-type systems are generally adaptable for the use of automatic control valves, because the normal practice is to maintain a controlled difference between the supply and return line pressures while varying the supply main line pressure in accordance with the heating load variations.

The extension of a water line above the trap is important to make certain that the condensate is cleared from the bottom rows of the coil. See Figure 5-5.

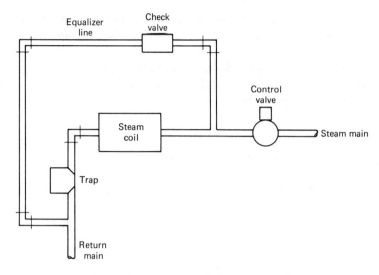

Figure 5-5 Steam-supplied air heating coil.

An equalizer line is also important in both atmospheric and vacuum return systems. This is an added protection to prevent the vacuum forming in the coil and keeping the water from draining into the trap.

STEAM TO HOT WATER CONVERTERS

Converter selection is very important. It is better to undersize them than to oversize them. The flash-type converters are very popular. Flash-type converters have a very small amount of water as compared to the volume of steam, and the leaving water temperature may be very close to the steam temperature. Because of this, the selection of the supply steam pressures and the valve pressure drop is very significant. It may be required that the converter deliver water at varying temperatures. When the difference between the temperature of the delivered water and the steam temperature goes down, the size and cost of the converter go up. As an example, a converter which supplies water at a temperature of 120°F would need to be only one-third the size of one which would supply the same building with water at 200°F. The capacity of a converter, unlike the coils for air treatment, will vary quite drastically with any changes in the steam pressure which is delivered to it. Let's consider a converter which supplies water to a building at 180°F. The converter is designed to operate with a 10-psi steam pressure delivered to it. Should the pressure inside the converter be reduced to 5 psi, there would be a 17% reduction in the capacity of the converter. If the steam pressure should be reduced to 2 psi, the converter would have only 70% of its design capacity. The steam supply main to a hot water converter should then be specified and selected. The control valves should also be designed so that the full design steam

Chap. 5   Pneumatic Control Valves   69

pressure can be obtained within the converter and still allow a sufficient valve pressure drop to provide the desired control. It is generally considered that an 80% pressure drop is to be desired; however, a 50% pressure drop may be used if the available steam main supply pressure prevents the use of an 80% pressure drop.

For example, if there is 10 lb of steam pressure available to the converter when an atmospheric return is used, an 80% valve pressure drop will leave 2 lb of steam in the converter. With only 2 lb of steam available, a large converter would be required. When these conditions are encountered, some type of compromise could be made by using a valve with a 50% pressure drop. In this situation, satisfactory control could be obtained; however, more expensive controls would need to be used.

STEAM HUMIDIFIERS

Some types of humidifiers in use today are water spray, steam grid, and the steam pan types. The steam grid is the most efficient to control. This is possible because steam is more readily diffused without the requirement of adding heat to the air which is being humidified and therefore the air receives very little sensible heat.

Pan-type humidifiers usually require 5 psi of steam pressure because the water must be made to boil in order to evaporate at a rate fast enough to be effective. Thus, the valve pressure drop must not be greater than the steam main line pressure less 5 psi. The spray and steam grid types of humidifiers require only 0.5 psi to 1 psi, respectively, for proper operation. The valve pressure drop is then determined by subtracting the humidifier inlet steam pressure from the supply main line pressure.

SELECTING STEAM VALVES

After the steam distribution system has been designed, and the steam pressures have been selected which will allow for effective operation of the control valve, the specific valve can also be selected. When selecting the valve, use the following steps:

1. Determine the amount of steam required in pounds per hour (W) in the following formula.
2. Determine the pressure drop across the valve.
3. Find the required capacity index (C_v).
4. Select the valve which has the required specifications.

The formulas which are given below can be used to determine the amount of steam per hour required for the different types of equipment:

1. When the Btu/hr is known (heat input), $W = $ Btu/hr $\div 1000$.
2. When the equivalent direct radiation (EDR) is known, $W = $ EDR $\times 0.24$.
3. When sizing the coil valves for forced air systems, $W = $ CFM $\times 1.08 \times TD_a/1000$, where CFM = air quantity in cubic feet per minute and TD_a = air temperature difference across the column.
4. When sizing the converter valves, that is, steam to hot water, $W = $ GPM $\times TD_w \times 0.49$, where GPM = water flow rated in gallons per minute and TD_w = water temperature difference.
5. When sizing steam jet humidifier valves, $W = $ CFM $\times 60 \times 0.75 \times (GR_1 - GR_2)/7000$, where CFM = air quantity in cubic feet per minute, GR_1 = grains of moisture per pound of air leaving the humidifier, and GR_2 = grains of moisture per pound of air entering the humidifier.
6. When sizing pan-type humidifier valves, $W = $ CFM $\times 60 \times (GR_1 - GR_2) \div 7000 \times 1.5$.

To determine the pressure drop across the valve, calculate as follows:

$$\text{pressure drop} = 80\% \; (P_1 - P_2)$$

where P_1 = steam supply main line pressure in psi and P_2 = steam return main line pressure in psi. $P_1 - P_2$ should equal the lowest pressure that will normally be possible across the supply and return lines nearest the control valve.

When all of the facts regarding the steam quantity and the pressure drop have been determined, the capacity index (C_v) required can also be determined. This value is determined by use of the valve sizing table. See Table 5-1.

When all of the information mentioned has been obtained, a valve can be selected which will meet the desired specifications. When selecting a valve, use the following general considerations:

1. When tight shutoff is required, use a single-seated valve. If this is not a requirement, a double-seated valve can be used to advantage because it will have a higher close-off rating.
2. The pressure drop limitations should be considered. These limitations are usually based on the normal life of the seat and disc when the steam is flowing through the valve at the maximum expected pressure drop. Because the greatest steam velocity and valve wear occur when the valve is almost closed off and since two-position valves are always opened or closed, the pressure drop limitations for the two-position valves are generally considered to be higher than those for modulation-type valves. The modulating type of valve may assume an almost closed position for long periods of time, resulting in greater wear on the valve and seat.
3. Consider the fluid temperature and pressure limitations. In this case either the valve body or the valve packing material may limit the amount of pressure and temperature that can be used in the system.

TABLE 5-1 VALVE SIZING TABLE

C_v	STEAM CAPACITY IN #/HR.												$W = 2.1 C_v \sqrt{\Delta P} \sqrt{P_1 + P_2}$	WATER CAPACITY GPM $Q = C_v \sqrt{\dfrac{\Delta P}{G}}$			
	5# STEAM @ PRESS. DIFF.			10# STEAM @ PRESS. DIFF.						15# STEAM @ PRESS. DIFF.				C_v	POUNDS DIFF. PRESSURE		
	2	3	4*	2	3	4	5	7	8*	5	8	10	12*		1	2	3	5
.25	4.38	5.4	6.25	4.6	5.68	6.59	7.39	8.69	9.3	7.84	9.92	11.1	12.12	.25	.354	.434	.56
.50	8.75	10.8	12.5	9.2	11.35	13.15	14.75	17.35	18.6	15.65	19.80	22.1	24.25	.50	.706	.865	1.10
.60	10.5	12.9	15.0	11.05	13.62	15.8	17.70	20.8	22.3	18.8	23.8	26.65	29.1	.60	.850	1.04	1.34
.80	14.0	17.3	20.0	14.72	18.15	21.1	23.6	27.8	29.8	25.1	31.7	35.6	38.8	.80	1.13	1.39	1.79
.90	15.8	19.5	22.5	16.56	20.4	23.7	26.6	31.3	33.5	28.2	35.7	39.8	43.6	.90	1.27	1.56	2.01
1.0	17.5	21.6	25.0	18.4	22.7	26.3	29.5	34.7	37.2	31.3	39.6	44.2	48.5	1.0	1.42	1.73	2.24
1.1	19.3	23.8	27.5	20.3	24.9	28.9	32.4	38.2	40.9	34.2	43.6	48.9	53.4	1.1	1.56	1.91	2.46
1.2	21.0	25.9	30.0	22.1	27.2	31.8	35.4	41.6	44.6	37.6	47.6	53.4	58.2	1.2	1.70	2.08	2.68
1.6	28.0	34.6	40.0	29.4	36.3	42.1	47.2	55.6	59.5	50.1	63.4	71.0	77.6	1.6	2.26	2.78	3.58
1.8	31.6	38.9	45.0	33.1	40.8	47.4	53.2	62.5	67.0	56.4	71.4	79.9	87.4	1.8	2.55	3.12	4.03
2.0	35.0	43.2	50.0	36.8	45.4	52.6	59.0	69.4	74.4	62.6	79.2	88.4	97.0	2.0	2.84	3.46	4.48
2.2	38.5	47.6	55.0	40.4	49.9	57.8	64.9	76.4	81.6	68.9	87.2	97.6	106.9	2.2	3.12	3.82	4.93
2.4	42.0	51.9	60.0	44.2	54.4	63.2	70.9	83.4	89.4	75.2	95.4	106.5	116.4	2.4	3.40	4.16	5.36
2.5	43.8	54.0	62.5	46.0	56.8	65.9	73.9	86.9	93.0	78.4	99.2	111.0	121.2	2.5	3.54	4.34	5.6
3.0	52.5	64.8	75.0	55.2	68.1	78.9	88.5	104.1	111.6	93.9	118.8	132.6	145.5	3.0	4.26	5.19	6.72
4.0	70.0	86.4	100.0	73.6	90.8	105.2	118.0	138.8	148.8	125.2	158.4	176.8	194.0	4.0	5.68	6.92	8.96
4.5	78.9	97.1	112.5	82.9	101	118.5	132.9	156.1	167.5	141.0	178	200	218	4.5	6.36	7.80	10.1
4.6	80.6	99.5	115.0	84.6	104.2	121	135.6	159.9	172	144.5	182	204.5	223	4.6	6.51	7.96	10.3
5.0	87.5	108.0	125.0	92.0	113.5	131.5	147.5	173.5	186	156.5	198	221	242.5	5.0	7.1	8.65	11.2
5.3	93.8	114.4	132.5	97.5	120.2	139.5	156.2	184	197	166	210	235	257	5.3	7.50	9.20	11.89
5.6	98.4	121.0	140.0	103.0	127.0	147.5	165.1	194.5	206	175.4	222	249	272	5.6	7.94	9.72	12.5
5.7	99.9	123.0	142.5	105.0	129.2	150	168	198	212	178.5	226	253	277	5.7	8.06	9.89	12.75
6.3	110.0	136.0	157.9	115.9	143.0	165.9	186	219	234	197.5	249.9	280	306	6.3	8.93	10.90	14.1
6.5	114.9	140.5	162.5	119.5	147.5	171	192	226	242	203.9	257.9	289	315	6.5	9.20	11.28	14.5
9.0	157.5	194.4	225.0	165.6	204.0	237	266	312.3	335	282	356.4	398	436.5	9.0	12.78	15.57	20.16
9.3	163.9	201.0	232.0	171.2	211	245	274	323	346	291.2	368	414	452	9.3	13.15	16.10	20.8
9.5	166.2	205.5	238.0	174.9	218	250	281	350	353	297.9	378	423	462	9.5	13.41	16.49	21.2
11	193	238.0	275.0	203	249	289	324	382	409	342	436	489	534	11	15.6	19.1	24.6
12	210	259	300	221	272	318	354	416	446	376	476	534	582	12	17.0	20.8	26.8
15	263	324	375	276	341	394	443	521	558	468	594	666	726	15	21.2	26.0	33.5
16	280	346	400	294	363	421	472	556	595	501	634	710	776	16	22.6	27.8	35.8
17	298	368	425	313	386	446	501	592	634	533	673	756	824	17	24.0	29.5	38.0
18	316	389	450	331	408	474	532	625	670	564	714	799	874	18	25.5	31.2	40.3
21	368	454	525	386	476	554	620	729	782	656	831	935	1019	21	29.97	36.4	47.0
22	385	476	550	404	499	578	649	764	816	689	872	976	1069	22	31.2	38.2	49.3
23	404	497	575	424	522	604	678	799	856	722	912	1021	1115	23	32.5	39.8	51.2
25	438	541	625	460	568	659	739	869	930	784	992	1110	1212	25	35.4	43.4	56.0
30	525	648	750	552	681	789	885	1041	1116	939	1188	1326	1455	30	42.6	51.9	67.2
38	666	820	950	699	636	998	1121	1320	1415	1190	1505	1689	1842	38	53.8	65.9	84.0
40	700	864	1000	736	908	1052	1180	1388	1488	1252	1584	1768	1940	40	56.8	69.2	89.6
47	824	1015	1175	865	1068	1239	1386	1632	1749	1472	1860	2065	2280	47	66.5	81.5	105.0
48	842	1038	1200	884	1089	1264	1419	1669	1785	1502	1900	2180	2330	48	67.9	83.4	107.2
70	1228	1512	1750	1288	1589	1841	2055	2429	2604	2191	2772	3094	3395	70	99.4	121.1	156.8
72	1260	1555	1800	1325	1635	1899	2125	2500	2680	2259	2850	3199	3490	72	102.0	124.7	161.0
75	1312	1620	1875	1380	1700	1975	2215	2610	2790	2350	2970	3330	3640	75	106.0	130.0	167.5
93	1639	2010	2320	1712	2110	2450	2740	3230	3460	2912	3680	4140	4520	93	131.5	161.0	208.0
94	1649	2030	2350	1730	2140	2470	2779	3270	3498	2945	3720	4170	4560	94	133	163	210
100	1750	2160	2500	1840	2270	2630	2950	3470	3720	3130	3960	4420	4850	100	142	173	224
102	1782	2210	2550	1865	2319	2680	3010	3540	3799	3199	4040	4530	4940	102	144	177	228
104	1820	2250	2600	1910	2360	2730	3070	3620	3870	3259	4120	4620	5040	104	147	180.5	232
164	2870	3540	4100	3020	3720	4320	4840	5700	6100	5140	6490	7270	7950	164	233.2	284	366
170	2980	3680	4250	3130	3860	4460	5010	5920	6340	5330	6730	7560	8740	170	240	295	380
174	Not Available for Steam			Not Available for Steam						Not Available for Steam				174	246	302	389
200	3500	4320	5000	3680	4540	5260	5900	6940	7440	6260	7920	8840	9700	200	284	346	448
250	Not Available for Steam			Not Available for Steam						Not Available for Steam				250	354	434	560
360	Not Available for Steam			Not Available for Steam						Not Available for Steam				360	509	624	805

*Recommended Pressure Drop (80% x P_1)

4. The flow characteristics should be considered. Normally use a linear or equal percentage relationship for the modulating-type valves and quick opening for two-position valves.
5. The ambient temperature of the actuator is important. Pneumatic-type actuators have a neoprene-type diaphragm which has lower ambient temperature ratings. Silicone rubber diaphragms are used for the higher actuator temperatures.
6. The needed body pattern should be determined. This includes such items as screwed, flanged, angle, straight through, three-way, etc.
7. Use the C_v requirement to select the proper valve. In some cases there are several different body styles which are available that have the same C_v capacity. Normally, however, the C_v rating will increase as the body size increases.
8. Select a valve which has the proper close-off rating. This is required so that it will close off against the highest pressure difference that will be encountered in that particular system.
9. A positive positioner may be required. This is especially so when the valve must be sequenced with other pieces of equipment. As an example, when a heating valve is used with a 4–6 psi operating pressure and a cooling valve with an 8–13 psi operating pressure, a positive positioner may be required if accurate control is required. In this case the positioner will allow the valve to move exactly as far as the controller signal demands, and the valve position will not be affected by the friction of the packing, the fluid pressures, etc.

WATER VALVES

The selection of the water valves and the design of the water system must be coordinated so as to accomplish the best and most accurate proportional control of the water system. When the original design is being considered, the location of all the control valves must be considered to ensure that the system will deliver the design flow at the full load conditions and not generate uncontrollable conditions when operating at the minimum load conditions. The selection of the valves based on the pressure differentials at the control valve locations will depend on the valve sizing at the full load conditions and the valve controllability and close off at the minimum load conditions.

There are three factors which affect the control valve pressure differentials:

1. The variations in the fluid flow
2. The type and size of the piping used in the distribution system
3. The pump characteristics and the regulation of the supply pressure differential

There are also three methods of providing control over a water system:

1. Control of the supply water temperature. This is the most effective method of controlling the Btu output of a water supply system.
2. Flow control is a suitable method of controlling the individual terminal units. The units are fan coils and induction-type units.
3. A combination of temperature and flow control of supply water. This is the best control method used for heating systems.

In practice there are several ways to vary the supply water temperature being fed to the coils. See Figure 5-6. In this type of system, the transmitter T_2 senses the outside air temperature. Transmitter T_1 senses the supply water temperature and maintains a preselected schedule of the water temperature according to the outside air temperature. Upon a drop in the outside air temperature, the supply water temperature is increased. Conversely, on a rise in the outside air temperature, the supply water temperature will decrease. Also, there is a reset schedule shown which is predetermined according to the design conditions and the system capacity.

When the flow control method is used to control the flow of water, the control of the Btu output of the supply water does not vary in direct proportion to the water flow. This is the simplest method used for providing control of individual terminal units which are fed from a single common heat source. The control of these units can be accomplished with either two-way valves or three-way coil bypass valves.

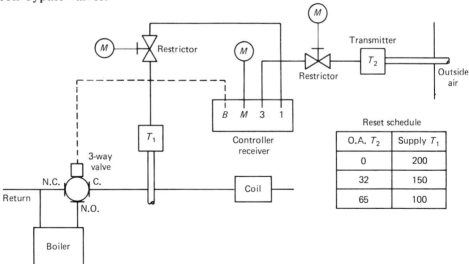

Figure 5-6 Discharge water temperature control system.

When using two-way valves for the control of the individual units, the flow of water through the supply system is varied, causing pressure variations which are dependent on the type and size of the distribution system. See Figure 5-7.

The use of three-way valves to control the flow to the coils is advantageous, especially when it is desired to maintain a constant flow of water through the system main lines. See Figure 5-8. In operation, as the valve reduces the water flow through the coil, it increases the flow through the bypass around the coil. In this manner, the total flow of water through the system is held relatively constant, and the friction loss through the system is also held relatively constant.

A deeper examination of the system will show that the total GPM which is circulated by the water pump does not always remain constant. When the system is at maximum demand, all of the water flows through the complete system, including the coil, and there is a friction loss equal to that imposed by the complete system piping. When the valve is in the minimum demand position, the water flows around the coil, reducing the total friction caused by the complete system. Thus, because there is usually less friction in the bypass circuit, there will be an increase in the flow of water through the system. When this situation occurs at only a few of the coils used in the system, a serious flow imbalance through the system can occur. If this condition occurs simultaneously at several coils through the complete circuit, the effect can be appreciable.

Therefore, the bypass circuit should be designed to offer the same amount of resistance as the circuit through the coil. When this step is omitted or is not practical, a balancing cock should be installed in the bypass circuit. See Figure 5-9. The balancing cock is used to add the necessary amount of resistance to the system.

In some cases, even when the circuits are balanced properly for the maximum and minimum conditions, the intermediate positions of the valve can possibly have a significant effect on the overall resistance of the complete system. For example, when one-half of the water is flowing through the coil while the other half is flowing through the bypass, the total friction loss through the circuit is considerably less than if all of the water was flowing through either of the circuits. This produces the same effect as when the valve is in either of the other extreme positions.

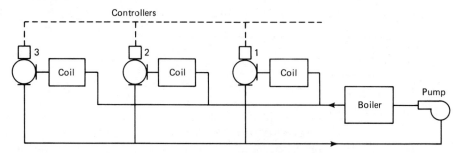

Figure 5-7 Water flow control using two-way valves.

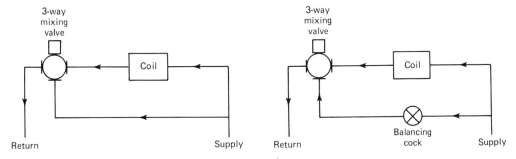

Figure 5-8 Coil by-pass using a three-way valve.

Figure 5-9 Coil by-pass using a balancing cock.

Even the use of three-way valves in conjunction with a properly designed and balanced system will not ensure that the system will be completely balanced under all operating conditions. Three-way valves, therefore, do not always cure all of the system's operating problems, and it is still necessary to properly design the system and to carefully consider all of the variables that could possibly affect the operation of the system. When using a three-way coil bypass valve, the following should be given serious consideration:

1. A three-way valve will control the coil output by varying the flow just the same as the two-way valve.
2. The piping costs can be more for a three-way than for a two-way valve system. This is especially true where the piping space is limited.
3. Three-way valves generally are available only with linear flow characteristics. The equal percentage two-way valves are usually better suited to control the flow of water to the coils where a close control is required or desired. This may be compensated for, however, to some degree by the use of scheduled supply water temperatures.
4. The constant flow of water in the system mains is possible when three-way valves are used without the use of added bypass valves.

The third method of system control by controlling both the temperature and the flow is advantageous except in large applications which use individual room temperature control because of the piping requirements to each individual heating unit. The most desirable method is to control the water flow to each heating unit by use of a thermostatically controlled valve. By the addition of a supply water temperature control system to the unit, individual room temperature control can be accomplished and even improved. See Figure 5-10.

This method of control has the following advantages:

1. When mild weather is experienced, the supply water temperature is reduced, which causes the individual heating unit control valves to open wider than when a constant water supply temperature is used. The wider-open valves allow for a more constant flow of water through the pump and the boiler

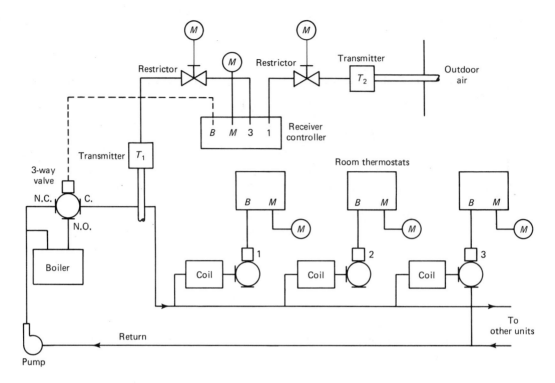

Figure 5-10 Temperature and flow control of supply water.

at a fairly constant volume at this time of reduced demand on the system.
2. During the initial warm-up period of a room or when the load is heavy, an overshooting of the temperature is not as apt to occur because the maximum heating unit output is regulated from the outdoor air temperature.
3. When the average conditions are experienced, the heating unit valves will be operating near the midstroke position rather than operating as two positions between the closed and the nearly closed position.

WATER CIRCULATING SYSTEMS

The main purpose of a water circulating system is to deliver the water to the coils at the system design flow and temperature when the system is under full demand without creating conditions which are uncontrollable when the system is at minimum demand conditions. The size of the circulating system in terms of pressure drop through the entire system will determine the maximum pressure buildup at the individual units when the system is under maximum load conditions. The type of circulating system in question will determine whether the pressure differential can be regulated in response to the load changes of the system.

Chap. 5 Pneumatic Control Valves   77

The two most common types of circulating systems are the direct return (see Figure 5-11) and the reverse return (see Figure 5-12). As can be seen in Figure 5-11, the first heating load to be taken from the main line is the first to be returned to the system. In these types of systems the pipe lengths are always unequal and are, therefore, unbalanced.

The pump in these types of systems must be sized to overcome the friction loss of the longest water circuit in the entire system. This presents a problem because the pressure head will be too high for the first part of the system, causing an excessive amount of water to flow through that part of the system, resulting in an uneven distribution of the heat and probable noisy operation. Due to the fact that the pump can only deliver a given amount of water for each amount of friction head pressure, the last part of the system will receive only a small amount of water and therefore a small amount of heat.

However, these types of systems can be balanced by placing balancing cocks in the water circuit to each of the heating units. Even when this practice is used, the system will remain in balance only when the flow of water through the system is constant. It can be seen that on a system which has several units the balancing process would be a time-consuming and a very expensive process. Because of this, the direct return system is recommended only for small, constant water flow types of systems.

When the reverse return type of circulating system is used, the first load to be taken from the system is the last to be returned to the system. This procedure produces a relatively constant water flow through the system because all of the circuits are approximately the same length (see Figure 5-12), thus almost completely eliminating the problem of system balance.

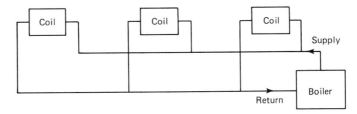

Figure 5-11 Two-pipe direct-return system.

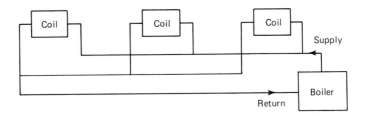

Figure 5-12 Two-pipe reverse-return system.

The reverse return type of system is the best of all the possible arrangements for maintaining the system balance and an almost constant pressure drop throughout the complete system. It should be noted, however, that unless some other type of provisions are used, there can possibly be a large increase in the pressure drop across the control valves as the flow of water to the heating units is reduced from maximum to zero.

SUPPLY PRESSURE DIFFERENTIAL REGULATION

Any decreases in the flow of supply water are accompanied by a decrease in the piping pressure loss and an increase in the pressure differential which is generated by the water pump. Any time a valve pressure drop at minimum water flow is more than three times the valve pressure drop at maximum water flow, some means of differential pressure regulation should be used to control the flow of water.

A constant pressure drop across the system in most applications is maintained by the use of a pump bypass. See Figure 5-13. As the control valves on each of the individual units decrease the flow of water through the system during a low load condition, the pressure across the supply and the return main lines increases. This increase in the pressure is too high to allow for the proper control by the unit valves.

More accurate control of the water flow can be achieved by the addition of a pump bypass and the use of a differential pressure regulation controller to monitor the pressure on either side of the pump and position the pump bypass valve. The differential pressure type of controller can be used to offset any gain in pressure because of the variations in the pump head. It cannot, however, com-

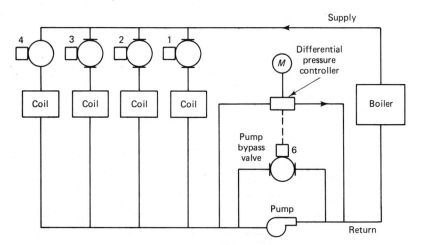

Figure 5-13   Pump by-pass system.

pensate for any pressure buildup because of piping design pressure loss. Because of this, the pump bypass is normally used in systems where the design piping pressure drop is less than two times the design unit valve drop.

Another popular method of regulating the system water pressure drop is by installing a bypass valve between the supply and the return main lines to allow the water to flow through the mains at a steady rate. Thus, the system frictional loss can be maintained at an almost constant rate. See Figure 5-14. The bypass valve should be set to open as the unit control valves close off the flow of water through the units. A differential water pressure controller is used to monitor the friction head across the supply and the return main lines. The controller modulates the system bypass valve to maintain a constant flow of water through the system.

WATER VALVE SELECTION

When the water circulating system has been designed and the system pressures have been determined, the control valves can be selected to provide the desired amount of water flow through the system.

When selecting the water valves, use the following steps:

1. Determine the gallons of water per minute (GPM) required.
2. Determine the desired pressure drop across each of the control valves.
3. Determine the capacity index (C_v) needed for proper operation.
4. Select the valves which provide the proper requirements.

When determining the GPM, it is necessary that the output Btu and the design water temperature drop of the heat exchanger be known. Use the following formula to determine the GPM:

$$GPM = Btu/hr \div K \times TD_w$$

where TD_w = water temperature difference and K = the factor which is selected from a table. See Table 5-2.

In cases where the equivalent direct radiation (EDR) is known, the following method is recommended:

$$GPM = EDR \times (correct\ Btu/hr \div EDR\ value\ K \times TD_w$$

See Table 5-3 regarding the EDR value.

An example for determining the GPM when the EDR is known is as follows: Determine the GPM required to pass through a control valve on a length of cast iron radiation which has an EDR rating of 150. The water temperature is 200°F entering and 180°F leaving:

$$GPM = EDR \times (correct\ Btu/hr/EDR) \div K \times TD_w$$
$$GPM = 150 \times 209 \div 484 \times 20$$

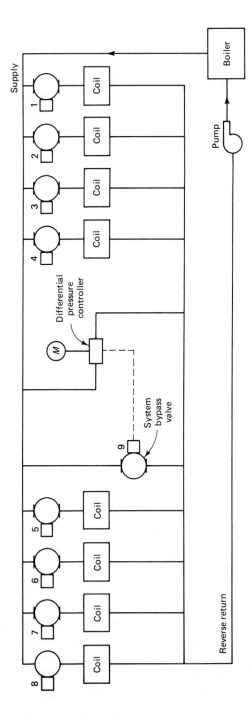

Figure 5-14 System by-pass valve location.

TABLE 5-2 K-FACTOR CHART

Water temperature (°F)	K	Water temperature (°F)	K
60	500	225	483
80	498	250	479
100	496	275	478
120	495	300	473
150	490	350	470
180	487	400	465
200	484		

TABLE 5-3 OUTPUT OF RADIATORS AND CONVECTORS

Average radiator or convector temperature (°F)	Cast iron radiator (Btu/hr/EOR[a])	Convector (Btu/hr/EOR[b])
215	240	240
200	209	205
190	187	183
180	167	162
170	148	140
160	129	120
150	111	102
140	93	85
130	76	69
120	60	53
110	45	39
100	31	27
90	18	16

[a] At 70°F room temperature.
[b] At 65°F inlet temperature.

$$\text{GPM} = 31{,}350 \div 9680$$

$$\text{GPM} = 3.23$$

If there is a design allowance of 2 psi of pressure drop across the control valve, the valve sizing table (Table 5-1) shows that a valve with a C_v of 3.0 would be recommended for the application being considered.

When calculating the GPM requirement for a hot water coil to be used in a fan coil installation, use the following formula:

$$\text{GPM} = \text{CFM} \times 1.08 \times TD_a \div K \times TD_w$$

where CFM = the air quantity in cubic feet per minute and TD_a = the air temperature difference.

When sizing the valves for a cold water installation, use the following formula:

$$\text{GPM} = \text{CFM} \times \text{Btu/lb of dry air removed} \div 113 \times TD_w$$

where dry air removed = total heat removed.

When this formula is being used, it is necessary to determine the number of pounds of dry air removed from an appropriate psychrometric chart.

The pressure drop across a water valve should be between 50 and 70% of the system design pressure difference between the supply and the return main lines closest to the valve. This pressure drop will provide the best modulating control of the water flow. As an example, should this pressure drop be 12 psi, then the desired pressure drop across the valve should be a minimum of 6 to 8.4 psi.

The final step in the selection of a water valve is to find a valve which most nearly meets all of the specifications. The basic considerations for selecting a water valve are as follows:

1. Determine the amount of water required per minute (GPM).
2. Determine the pressure drop across the valve.
3. Find the required capacity index (C_v).
4. Select the valve which meets the required specifications.

REVIEW QUESTIONS

1. What is the part of the automatic control valve which causes the valve stem to move?
2. Define the valve disc.
3. What are the classifications of pneumatic control valves?
4. What are the control actions of the valve called?
5. Which type of valve requires the most force to close it?
6. Define the capacity index of a valve.
7. Define the close-off rating of a valve.
8. What is the rangeability rating of a valve?
9. What is the purpose of having a dead spot between two different valve spring pressures when used in cooling/heating applications?
10. What is the first step in selecting the steam pressures in a steam heating system?
11. What should the valve pressure drop be in a modulating-type system?
12. Why should the stem return main line pressure remain constant if at all possible in a steam heating system?
13. When selecting a steam to hot water converter, is it better to undersize or oversize the converter?
14. Write the formula for sizing coil valves for forced air systems.
15. Name three methods of controlling a water system.
16. What is the simplest method used for providing control of individual terminal units which are fed from a single common heat source?
17. How should a coil bypass circuit be designed?
18. What are the two most common types of water circulating systems?
19. How should the pump be sized in a water system?
20. In what type of system is the problem of system balancing almost eliminated?
21. When should some means of differential pressure regulation be used to control the flow of water through the system?
22. Where is the differential pressure type of controller used?
23. Write the formula for calculating the flow of water required in a circuit.
24. Write the formula for calculating the GPM requirement of a coil.
25. What should the pressure drop across a water valve be?

6

Pneumatic Relays

Pneumatic relays are used in control systems to provide a large variety of switching functions in the pneumatic control system. The use of these relays in the pneumatic control systems is almost unending. We will discuss a few of the most common uses to provide a basic understanding of some of the more popular relays.

DIVERTING RELAYS

Diverting relays are three-way air valves which are designed to provide a wide variety of switching and interlocking functions in the pneumatic control system. Their primary use is to convert a pneumatic signal, at a predetermined setting, into a signal for a final control device. The flexibility and the design of the diverting relays allow them to be piped normally to provide a great number of functions such as to feed and exhaust a branch air line or to divert one supply line to either one of two supply lines or to one branch line. See Figure 6-1. In this application the relay is used for diverting either summer or winter operating air pressures to the summer/winter thermostats. This application uses an outside air transmitter to provide the switching signal to the diverting relay. The outside air transmitter used in this example has a range of 0–100 °F, and the desired switching temperature is 65 °F outside temperature. When the outside air temperature is at 65 °F, the transmitter used in this example will have an output signal of 10.8 psi. The diverting relay then is set to switch at 10.8 psi of air pressure. The winter main line operating pressure is directed to the normally open port of the diverting relay and the summer operating air pressure to the normally closed port. The outside air transmitter is a direct-acting relay, and a signal from the transmitter

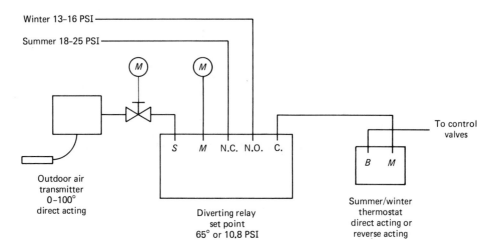

Figure 6-1 Typical diverting relay application.

above the 10.8-psi set point will connect the normally closed and the common ports, while a signal below the set point will connect the normally closed and the common ports. The common port serves as the main air connection to the summer/winter thermostats. The branch signal from the summer/winter thermostats will be directed to the final control device.

REVERSING RELAYS

The reversing relay is designed to be used in pneumatic control systems which require a reversing signal from the controlling device. The output line pressure of the reversing relay decreases in a direct proportion to an increase in the input signal pressure.

For a typical way in which the reversing relay can be used, see Figure 6-2. In this type of application the reverse-acting space thermostat is used to control both the heating and the cooling valves. The valves are of the normally closed type. The branch line signal from the thermostat is directed to the heating valve and then to the reversing relay. The signal is directed from the reversing relay to the cooling valve. The main air signal is also piped into the reversing relay. As the temperature inside the space drops, the branch line pressure from the thermostat is increased, thus causing the heating valve to open, allowing hot water to the coil. This same signal also goes to the reversing relay where it is reversed to a decrease in the pressure to the cooling valve. This reversed pressure causes the cooling valve to close. As an increase in the space temperature is sensed by the thermostat, its output pressure decreases, closing the heating valve. Since the signal to the reversing relay is reversed, its output pressure is increased and causes the cooling valve to open.

Chap. 6   Pneumatic Relays

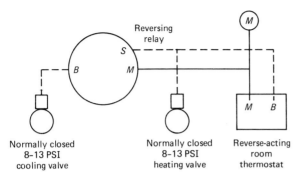

Figure 6-2 Typical reversing relay application.

PNEUMATIC-ELECTRIC RELAYS

Pneumatic-electric relays, sometimes referred to as PE relays, are used in pneumatic control systems which require a conversion of gradual air pressure change to a positive electrical switching action in the system. Some of the more popular applications of these relays are the starting and stopping of electric motors used on ventilating systems, fan coil units, air handling units, unit fans, unit heaters, water chillers, and hot water pumps. There are several different models which are available with the main difference being in the switching action and the type of differential adjustment. Most of these types of relays use the single-pole double-throw snap-acting switching arrangement. Some manufacturers incorporate a double-pole single-throw switching arrangement which will either open or close on a drop or a rise in system pressure. Some types are available with a special narrow differential. Also available is a switch having two single-pole double-throw contacts, with separate differential adjustments for each switch.

In operation, a change in the pressure signal from the outside air transmitter is the force used to open or close the PE switch contacts, which in turn either energize or de-energize an electrical circuit. See Figure 6-3. Often there are separate PE switches used to control the operation of the chilled water pump starter and the hot water pump starter. Each switch is adjusted so that the contacts either open or close at a predetermined pressure signal from the outside air transmitter. The PE switch which is used in our example is a single-pole double-throw type of switch, and it can be wired to provide either normally open or normally closed operation. In our example, the starter to the chilled water pump would be connected to the open common terminal of the PE switch. On an increase in the pressure signal from the outside air transmitter the contacts will close and energize the pump starter. The hot water starter is connected to the common and closed terminals of the PE switch. As the signal from the outside air transmitter decreases below the set point of the PE switch, the contacts will make the electric circuit and energize the hot water pump starter.

Another type of PE switch which is available and is used to control the opera-

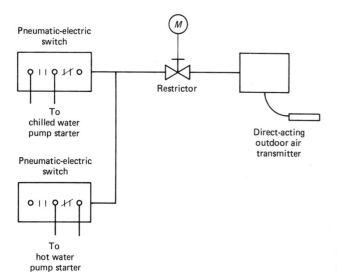

Figure 6-3 Typical pneumatic/electric switch connections.

tion of a unit fan and an electric-pneumatic relay is described in Chapter 4 under the heading "Day/Night System."

ELECTRIC-PNEUMATIC RELAYS

Electric-pneumatic relays use a solenoid type of operation to provide a two-position action. There are three connections provided which are marked normally open, normally closed, and common. They are especially useful in applications which use an electrical circuit to control a pneumatically operated device. They may be used to direct the supply air to a pneumatic device when the coil is energized, depending on the supply and the exhaust air connections.

A typical application is when the supply air is connected to the relays normally closed port and the control device is connected to the common port. Upon energizing the solenoid, the normally closed port will open, and the supply air is directed to the control device. When the solenoid is de-energized, the supply air connection is closed, and the normally open port allows the air to escape from the control device. To reverse the action of the relay, connect the supply air to the normally open port and use the normally closed port to exhaust the air from the control device.

An example of the use of this type of relay would be when an electric-pneumatic relay is used to energize a pneumatic-electric relay. See Figure 6-4. When it is energized, it allows the branch air from the controller to close the outside air damper and the heating control valve. When it is de-energized, the electric-pneumatic relay is closed and exhausts the air to the damper actuator and the heating control valve out the normally open relay port.

Another example using this relay is shown in Figure 6-5. When the unit

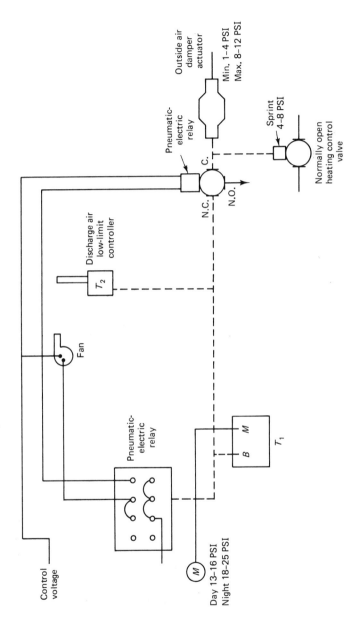

Figure 6-4 Day/night ventilator cycle.

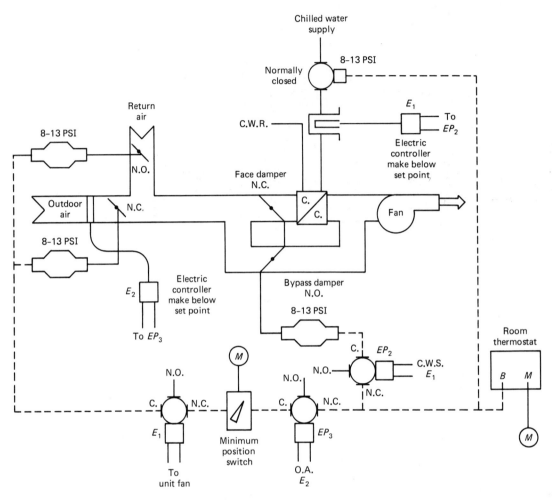

Figure 6-5 Electric pneumatic relay functions.

fan is started, EP_1 is energized, and the outside air damper opens to a minimum position which is determined by the setting of the minimum positioning switch. As the fan stops, EP_1 is de-energized, and the outside air dampers return to their normal positions. When chilled water is available, the electric temperature controller E_1 energizes EP_2, permitting air from the room thermostat to modulate the chilled water valve and the face and bypass dampers in sequence. If the chilled water is not available, the face damper closes, and the bypass damper opens. When the outside air temperature is below 60°F, the electric temperature controller E_2 is sensing the outside air temperature and energizes EP_3, allowing the air from the room controller to modulate the outside air and the return air dampers. When the outside air temperature is above 60°F, EP_3 is de-energized, and the unit fan is in operation. The outside air damper goes to its minimum position, which is determined by the minimum positioning switch.

PRESSURE SELECTOR RELAYS

Pressure selector relays are used in pneumatic control systems in which the application requires the comparison, selection, and transmission of one or two or more proportional signals supplied to the pressure selector relay. These units are available as either low-pressure selectors or high-pressure selectors. The low-pressure selectors receive the input signals and compare, select, and transmit the lower of these pressures. The high-pressure selectors receive the input signals and transmit the higher signal.

The high-pressure selector is usually used on cooling applications because either the thermostat or the humidistat can control the cooling valve. See Figure 6-6. When both the humidistat and the thermostat are calling for operation of

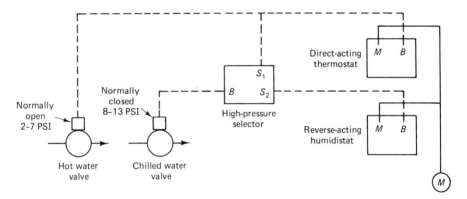

Figure 6-6 Typical high-pressure selector relay.

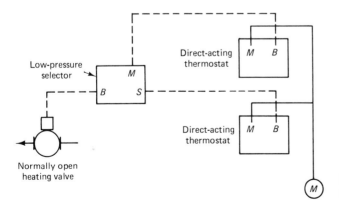

Figure 6-7 Typical low-pressure selector relay connections.

the cooling valve, the controller which has the stronger signal will take preference for system operation. In this illustration, the thermostat has direct control over the heating valve.

For a typical low-pressure application, see Figure 6-7. In this example, the signals from two separate thermostats are compared, and the lower of the two signals is transmitted to the heating valve.

AIR MOTION RELAYS

The purpose of air motion relays is to sense the suction and/or the discharge air pressures across a coil or a fan and to control the pneumatic controllers which are piped downstream from them. The sensing lines are located at the fan suction and discharge and are piped to the high- and low-pressure ports of the relay. In this manner the relay is able to detect whether or not the fan is operating.

Also, the air motion relay can be used to provide the same function as the electric-pneumatic relay when used on the outside air damper in a standard unit ventilating cycle. In this manner, the air motion relay will allow the outside air damper to operate only when the system fan is in operation. If an electric-pneumatic relay is used in the control system, the fan circuit can be energized, and the outside air damper will be operable even when the fan motor will not operate. When an air motion relay is used, the fan motor must be in proper operating condition so that the proper amount of pressure can be created to actuate the relay. When an air motion relay is used, the electric wiring to the electric-pneumatic relay is eliminated along with the possibility of the noise or the hum which sometimes accompanies relays that are electrically actuated.

When the air motion relay is to be used, there must be sufficient air pressure differential across the fan to cause the relay to operate as designed. See Figure 6-8. When this relay is used in this manner, the low-pressure sensing port is used

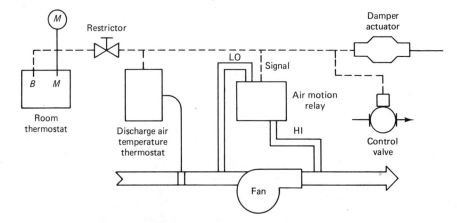

Figure 6-8 Air motion relay as used in a unit ventilator application.

as a reference port. In some cases, it may be necessary to pipe the high-pressure port so that it will sense the fan discharge pressure if there is not a sufficient amount of air pressure differential between the fan suction pressure and the reference air pressure.

During operation of the unit fan, the air motion relay senses the air pressure differential across the fan, which causes the signal port to close. The closing of this port will allow the air pressure downstream from the device to build up by using either a room thermostat or a discharge thermostat to control the pneumatic valve and damper actuator. At times when the unit fan is not operating, the equalized air pressure across the relay diaphragm will allow the diaphragm to fall away from the signal port. The air pressure remaining at the signal port will then exhaust to the atmosphere through the low-pressure sensing port.

AVERAGING RELAYS

The averaging relay is designed to be used in pneumatic control systems where the particular application requires operation of a final control device by the average signal from two other pneumatic controllers in the system. See Figure 6-9. In this example, the two room thermostats send their respective branch signals to the averaging relay. The relay then transmits the average pressure of the two signals to control the heating water valve.

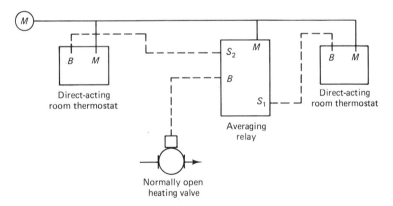

Figure 6-9 Typical averaging relay connections.

PNEUMATIC SWITCHES

The two most popular pneumatic switches are the gradual switch and the minimum positioning switch (accumulator). The following is a discussion of these two devices.

Gradual Switch

The gradual switch is a manually operated switch which is designed to deliver a preselected amount of air pressure from 0 to 20 psi to its connecting branch line. This switch is popular in pneumatic control systems for the remote positioning of the final control devices and also as a remote control point adjuster for receiver controllers or submaster controllers. See Figure 6-10. In this example, the gradual switch is used to provide a remote type of manual adjustment to the receiver controller. When the system operation is being monitored by a central control panel, it may be desirable to either raise or lower the set point of the receiver controller due to some type of unusual demand situation. In this instance, the result of lowering or raising the air pressure signal from the gradual switch to the receiver controller would be to vary the output signal of the receiver controller to the control valve.

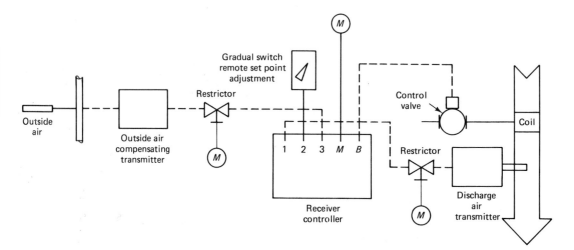

Figure 6-10 Typical gradual switch connections.

Minimum Positioning Switch (Accumulator)

The minimum positioning switch is a device which is commonly used in economizer types of systems. The main purpose of these types of systems is to use the outside air for cooling when the outside air temperature is low enough to allow this type of operation. See Figure 6-11. In this illustration the minimum positioning switch is used to position the outside air and the return air dampers.

In operation, the controls are actuated when the unit fan is in operation by the electric-pneumatic relay. The controller T_1 controls the operation of the dampers in order to maintain a mixed air temperature of 65°F. When the outside air temperature drops, the outside air dampers are moved further toward

Chap. 6   Pneumatic Relays   93

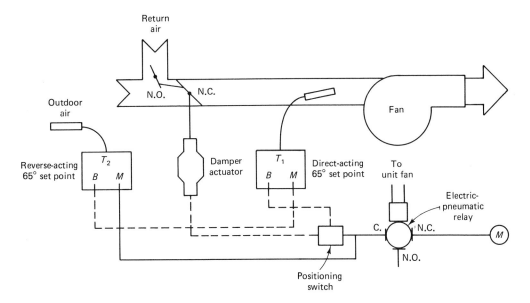

Figure 6-11   Typical minimum-positioning switch connections.

the fully closed position. The minimum positioning switch is used to prevent fully closing the outside air dampers and shutting off all the ventilation air to the building. The action of this switch is to allow air to pass to the damper actuator at whatever pressure is required by the controller T_1 until the required pressure is below 2 psi. At this time, the minimum positioning switch takes control and maintains 2 psi of air pressure in the line to the damper. This function prevents the operator from fully closing the outside air damper.

Should the outside air temperature rise above 65 °F, the temperature controller labeled T_2 will take control of the damper actuator. The damper is then moved to a predetermined position in relation to the outside air temperature as determined by the switch settings.

When the unit fan motor stops, air is exhausted through the normally open port of the electric-pneumatic relay. The damper actuator then returns to its original position, which closes the outside air damper.

REVIEW QUESTIONS

1. What is the purpose of relays in a pneumatic control system?
2. What are diverting relays?
3. What is the purpose of diverting relays?
4. When are pneumatic-electric relays used?
5. Where are electric-pneumatic relays used?

6. Where are selector relays used?
7. What is the purpose of air motion relays?
8. Where is the averaging relay designed to be used?
9. What is the gradual switch designed to do?
10. What is the purpose of the minimum positioning switch?
11. On the diverting relay, what port serves as the main air connection to the thermostat?
12. For what purpose is the reversing relay used?
13. To what does the output line pressure of a reversing relay decrease?
14. What are the main differences between the different models of PE relays?
15. From where is the force obtained to operate the PE relay?
16. What type of action is provided by an EP relay?
17. What pressure does the low-pressure selector relay transmit?
18. In what applications is the high-pressure selector relay used?
19. On a unit fan application, what pressure does the air motion relay sense?
20. What are the two most popular pneumatic switches?

7

Fundamentals of Electronic Control Systems

This chapter is a description of the electronic control system. Electronic equipment is different from the electric type because the electronic equipment provides proportional-type control with resistance sensing elements, while electric equipment is basically two-position in function, having contacts which are opened and closed by the sensing elements. Also in this chapter, the fundamentals of electricity will be discussed, with the presented principles being applied to the electronic control equipment.

BASIC ELECTRICITY

The following theories are all a part of the principles which are involved in the study of basic electricity. They are important for the understanding of the basic theories of electronic control systems and how they operate.

Ohm's Law

For any electrical circuit to perform a useful service, it must contain three basic elements: current, voltage, and resistance.

The source of electrical energy that we are all very familiar with is the battery. The battery in an electrical circuit performs the same function as the pump in a water circuit. The pump produces a water pressure which is measured in pounds per square inch (psi), and the battery produces an electromotive force (emf) which is measured in volts.

The amount of water which is flowing (GPM) through a water circuit is determined by the size of the water pump, the amount of pressure that it produces, and the amount of restriction presented by the circuit. In the same sense, the amount of electrical current (amperes or amps) which flows through an electric circuit depends on the voltage of the battery and the amount of resistance (ohms) which is offered by the electrical circuit. The voltage is the force which pushes the current through the circuit, the resistance is the opposition to the force, and the current is the amount of flow or the movement of the electrons through the circuit. The current is the flow rate or the number of electrons which pass a given point in the wire in 1 sec. One volt is required to force 1 A of current through 1 Ω of resistance. Mathematically this is expressed as E (voltage) $= I$ (current in amperes) $\times R$ (resistance in ohms). All of the electric and electronic theories are based on this formula, and it is called Ohm's law. The components can be rearranged as required to find any one of the various factors as follows: $E = IR$; $I = E/R$; $R = E/I$.

As an example we can connect a light bulb across a 12-V battery. See Figure 7-1. The light bulb offers 1 Ω of resistance. Determine the amount of current flowing through the circuit:

$$I \text{ (current)} = \frac{E \text{ (voltage)}}{R \text{ (resistance)}}$$

$$I = \frac{12 \text{ (volts)}}{1 \text{ (ohm)}} = 12$$

Therefore, a current of 12 A is flowing through the circuit and the light bulb.

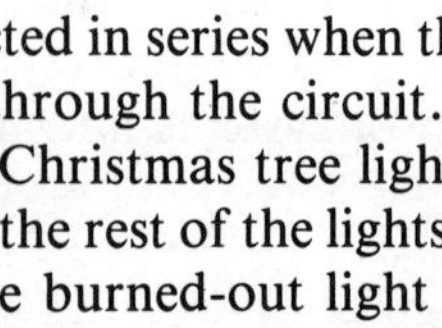

Figure 7-1 Calculating current flow through a circuit.

Series Connections

Two resistances are connected in series when the same current must flow through both loads when passing through the circuit. See Figure 7-2. A good example of a series circuit is when Christmas tree lights are connected in series and one of the lights burns out and the rest of the lights also go out. This happens because the circuit is broken in the burned-out light bulb.

The total resistance in a series circuit is equal to the sum of all the individual resistances in the circuit:

$$R_t = R_1 + R_2 + R_3 \cdots$$

In the example shown in Figure 7-2, $R_t = 4 \, \Omega + 1 \, \Omega = 5 \, \Omega$ of total resistance.

To calculate the amount of current flowing through the circuit,

$$I = \frac{E}{R} = \frac{10 \text{ V}}{5 \, \Omega} = 2 \text{ A}$$

A current of 2 A is flowing through this circuit.

In a series circuit, the current is the same through each one of the resistances. However, the voltage drop across each resistance is dependent on the amount of resistance offered by the resistance. When the current through a resistance and the amount of resistance offered by the load are known, Ohm's law can be applied to determine the voltage drop across the resistance:

$$E = IR = 2 \text{ A} \times 1 \text{ }\Omega = 2 \text{ V}$$

Thus, a 1-Ω resistance will have 2 V dropped across it if the applied voltage is 10 V. The remainder of the 10 V must be dropped across the 4-Ω resistance:

$$E = 2 \text{ A} \times 4 \text{ }\Omega = 8 \text{ V}$$

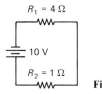

Figure 7-2 A series circuit.

Parallel Connections

In a parallel circuit, the voltage drop across each of the branch lines will always be the same, and the current flow through each resistance will depend on the value of the resistance. See Figure 7-3.

$$I_1 = \frac{E}{R_1} = \frac{10}{4} = 2.5 \text{ A}$$

$$I_2 = \frac{E}{R_2} = \frac{10}{1} = 10.0 \text{ A}$$

Since both of the bulbs draw current from the same battery, the current supplied is equal to the current flow through both bulbs:

$$I_t = I_1 + I_2 = 2.5 = 10 = 12.5 \text{ A}$$

When Christmas tree lights are connected in parallel and one of the bulbs burns out, the remainder of the bulbs will not be affected. This is possible because each of the bulbs has its own connection to the source of electricity.

The total resistance of a parallel circuit is always less than the smallest resistance in the circuit. The total resistance for a parallel circuit can be calculated with the following formula:

$$\frac{1}{R_t} = \frac{1}{R_1} + \frac{1}{R_2} + \frac{1}{R_3} + \cdots$$

The total current flowing through a parallel circuit is equal to the sum of the individual currents in the circuit:

$$I_t = I_1 + I_2 + I_3 + \cdots$$

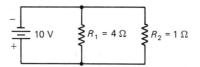

Figure 7-3 A parallel circuit.

When we apply Ohm's law, $I = E/R$, we have

$$\frac{E_t}{R_t} = \frac{E_1}{R_1} + \frac{E_2}{R_2} + \frac{E_3}{R_3} + \cdots$$

The voltage drop across each of the resistances is the same:

$$E_t = E_1 = E_2 = E_3 = \cdots$$

When this is substituted into the current equation, we have

$$\frac{E_t}{R_t} = \frac{E_t}{R_1} + \frac{E_t}{R_2} + \frac{E_t}{R_3} + \cdots$$

ac and dc Power

There are two different forms of electrical energy, with the major difference between them being the characteristics of the current flow. A dc power source supplies electricity which flows in only one direction. An ac power source supplies electric current which flows alternately in one direction and then in the other.

The batteries and the generators which are used in automobiles are a good example of dc electrical power sources. A dc current flow is, generally, thought of as a continuous unidirectional flow that is constant in its magnitude. However, in some instances a current which changes in magnitude but not in direction is also considered to be dc power.

The power which is supplied to us by the power companies is a common example of an ac power source. When the magnitude of the current is recorded as it varies with the time, the shape of the resulting curve is known as the waveform. The waveform which is produced by the power companies is known as the sine wave. See Figure 7-4.

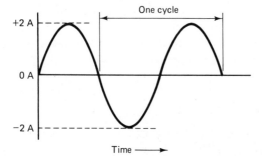

Figure 7-4 Sine wave of AC current.

Chap. 7   Fundamentals of Electronic Control Systems   99

When an ac current passes through a complete set of positive and negative values, as shown on the waveform, it has gone through a complete cycle. The frequency of ac current or voltage is determined by the number of cycles that are completed in 1 sec. The frequency of the current generated in the United States is 60 cycles/sec, known as 60 Hz.

All of the calculations made when considering ac current are based on sine waves. There are four values of the sine wave that we are concerned with:

1. *Instantaneous value*: The voltage or the current in an ac circuit is constantly changing. The value will vary from zero to a maximum and then back to zero again. If the value is measured at any given instant, the instantaneous value of the current or voltage will be given.
2. *Maximum value*: For two brief instances in each of the cycles that the sine wave travels, a maximum value will occur; one of the maximums will be negative, and the other will be a positive value. These values are often referred to as the peak values. The two terms have identical meanings and are used interchangeably.
3. *Average value*: Because the negative and the positive halves of the sine wave are identical, the average value can be found by determining the area located below the wave and by calculating what the dc value would be that would enclose the same area for the same period of time. When calculating either the positive or the negative half of the sine wave, the average value is 0.636 times the maximum value indicated by the waveform.
4. *Effective value* (rms): The effective value of an ac voltage or current is equal to a dc voltage or current which is required to provide the same average heating effect.

The heating effect is always independent of the direction of the current flow and is, therefore, the same for a full sine-wave cycle as it is for a half cycle of a sine wave. The effective value of an ac sine wave is equal to 0.707 times the maximum value. Therefore, an ac current sine wave which has a maximum value of 14 A will produce the same amount of heat as 10 A in a dc circuit. The heating effect will vary according to the square of the voltage or the current. If the instantaneous values of the current or the voltage are squared, we will obtain a waveform that is proportional to the instantaneous power, or the heating effect of the original sine wave. The average value of this new waveform is representative of the average power that will be supplied. The square root of this average value represents the voltage or the current value that represents the heating effect of the original sine wave of voltage or current. This figure represents the effective value of the sine wave, or the rms value, the square root of the average of the squared waveform.

The effective values of the voltage and the current are more important and

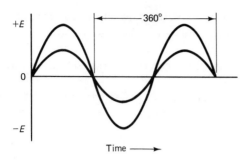

Figure 7-5 Sine waves in phase but unequal in amplitude.

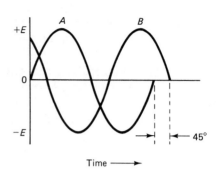

Figure 7-6 Sine waves 45 degrees out of phase.

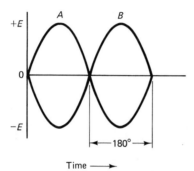

Figure 7-7 Sine waves 180 degrees out of phase.

useful than the instantaneous, maximum, or average values. Most of the ac ammeters and voltmeters are calibrated in rms values.

The phase of an ac voltage is in reference to the relationship of its instantaneous polarity to that of another ac voltage. Two sine waves may be in phase but have different values. See Figure 7-5. Figure 7-6 shows two sine waves which are 45° out of phase. Figure 7-7 shows two sine waves which are 180° out of phase.

The length of a sine wave is measured in angular degrees because each cycle is a repetition of the previous waveform. One complete cycle of a sine wave is 360 angular degrees.

Power in dc Circuits

Work is performed anytime that a force causes some motion. The force which causes the movement of electricity is called voltage. Energy is expended when voltage causes a movement of electrons from one point to another. The rate of doing electrical work, or the rate of transforming electrical energy, is usually expressed in watts or kilowatts. A kilowatt is 1000 W. When electricity is doing work in a dc circuit, 1 V forcing a current of 1 A through a resistance of 1Ω

Chap. 7   Fundamentals of Electronic Control Systems   101

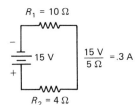

Figure 7-8 Calculating current in a series circuit.

represents 1 W of electrical power being used. To calculate power, use the following formula:

$$P \text{ (watts)} = E \text{ (volts)} \times I \text{ (amperes)}$$

Example: Find the power used by the resistances in a series circuit; use the values shown in Figure 7-8.

$$I = \frac{E}{R_t} = \frac{15}{5} = 3 \text{ A}$$

Then

$$P = E \times I$$
$$P = 15 \text{ V} \times 3 \text{ A} = 45 \text{ W}$$

This same method can be used in calculating the power used in a parallel circuit. See Figure 7-9. First find the total circuit current:

$$I = \frac{E}{R_t} = \frac{15}{0.8} = 18.75 \text{ A}$$

Now, using the power formula, we obtain

$$P = E \times I = 15 \text{ V} \times 18.75 \text{ A} = 281.25 \text{ W}$$

The power which is used in a circuit can be calculated if any two of the three values of current, voltage, and resistance are known: When the current is known,

$$P = \frac{E(E)}{R} = \frac{E^2}{R}$$

When the voltage is known,

$$P = (IR) I = I^2 R$$

When the resistance is known,

$$P = EI$$

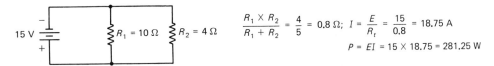

Figure 7-9 Calculating current and power in a parallel circuit.

Power Rating of Equipment

In most instances electrical equipment is rated in both voltage and power (voltage and watts). Most of the more common types of equipment, such as light bulbs, are rated in wattage because the voltage is generally taken as being 120 V. The wattage rating of a piece of equipment is an indication of the amount of current which will flow through the circuit. The greater the amount of current flowing through the circuit, the greater the amount of work that is being done.

The power rating of electric motors, resistors, and other electrical equipment indicates the rate at which these devices are to convert the electrical energy into some other form of energy such as heat or light. Whenever this wattage rating is exceeded, the excess amount of energy is usually converted to heat, and the equipment will overheat and in some cases it will be damaged beyond repair.

Electrical resistors are rated in the amount of watts that may dissipate along with the ohm rating. Resistors which have the same resistance value are available with different wattage ratings. Carbon composition or ceramic resistors are generally rated from approximately $\frac{1}{10}$ to 2 W. When higher wattage ratings are required, wire-wound resistors are generally used. In most cases, the larger the physical size of a resistor, the higher its wattage rating, because the larger amount of material will absorb and give off more heat more easily.

Inductance

Inductance is a factor that must be considered in electrical circuits. Inductance is that property of an electrical circuit, or a circuit component, which opposes any change in the current flowing through it. When there is any attempt to change the current flowing through the inductor, a self-induced voltage is generated in the coil (inductor). The term coil and inductor are often used interchangeably. An electrical coil is a series of rings or a spiral of wire wound around a core of some type. The polarity of the self-induced voltage is opposite to the voltage from the power source and tends to oppose any change in the current which generated it. Inductance, therefore, can be compared to kinetic energy in a mechanical system tending to keep a body moving at a constant velocity. See Figure 7-10.

Figure 7-10 Inductor circuit.

Capacitance

Capacitance is a physical property of the electrical circuit, just as resistance is. The capacitance of an electrical circuit is that property that opposes any change in the voltage across it. A capacitor in an electrical circuit allows electrons to

Figure 7-11 Capacitive circuit.

be stored in it the same as a gas or a liquid can be stored in a container. The capacity of the container is rated in gallons or in cubic feet, while the capacity of a capacitor is rated in farads or microfarads. The farad is a unit of capacitance. A capacitor has a rating of 1 F when a voltage change of 1 V/sec across its terminals produces a displacement of 1 A of current flow. See Figure 7-11.

Capacitors are rated for only a small portion of a farad because a farad is very large. Their capacity is either rated in microfarads (millionths of a farad) or picofarads (millionths of a microfarad).

There are two conducting surfaces in a capacitor. These surfaces are made from materials which have a very low resistance to the flow of electricity. These surfaces are separated by a material called a dielectric. A dielectric is an insulating material having an almost infinite resistance to the flow of electricity.

There are three purposes for using capacitors in an electronic circuit:

1. To bypass or filter out the ac component of a complex waveform
2. To couple an ac signal from one section of a circuit to another
3. To block out and/or stabilize any dc potential from some component

Power in ac Circuits

When Ohm's law is being considered for ac circuits, it states that the "current flowing through an ac circuit equals the voltage which is impressed across that circuit divided by the impedance of the circuit." This is the same as Ohm's law for dc circuits with the exception that the word *impedance* is substituted for *resistance*. The impedance of a circuit is the total opposition to the current flow through an ac circuit which is offered by the total resistance offered by the resistance, capacitance, and inductance. The ohm is the unit of impedance in an ac circuit. Ohm's law formulas for ac circuits are

$$I = \frac{E}{Z} \quad (Z = \text{impedance})$$

$$Z = \frac{E}{I}$$

$$E = IZ$$

The power which is used in a purely resistive ac circuit is calculated in the same manner as that in a dc circuit, $P = EI$.

When capacitance or inductance is added to a circuit, the formula $P = EI$ does not indicate the actual wattage being used by the circuit. In a circuit where the capacitance is greater than the inductance or the resistance, the current will lead the voltage. In a circuit where the inductance is greater than the capacitance and the resistance, the current will lag the voltage. The phase difference between the voltage and the current is known as the *power factor* and is equal to the cosine of the shared angle. When the current and voltage have a large phase angle between them, the power factor is said to be low. When the current and voltage are almost in phase with each other, the power factor is said to be high. When the angle between the voltage and the current is 90°, the power factor is said to be zero. When the voltage and current are exactly in phase, the power factor is said to be unity. The higher the power factor in ac circuits, the lower the circuit losses. The inductive and capacitive components in an ac circuit do not use any power from the power source; however, they do draw current from the power source. This extra curent flow will cause losses in the circuit, the generator, and the wires which are used to make up the system. This increased current flow may also require larger system components than would be required with a higher power factor. Because of these factors, the power company will in some cases penalize a customer using a low power factor load by increasing the cost of the electricity.

When two ac motors have the same horsepower rating, the electrical power consumed by both motors will be the same when they are doing the same amount of mechanical work. In cases where one motor has a higher power factor than the other, it will require less current to do the same amount of work. For example, if we have two motors, one with a power factor of 0.9 and the other with a power factor of 0.7, and both are rated 1 hp (746 W) and 120 V ac,

$$P = EI \times \text{power factor} = 746$$
$$746 = 120 \times I \times 0.9$$
$$I = 6.92 \text{ A}$$
$$P = EI \times \text{power factor} = 746$$
$$746 = 120 \times I \times 0.7$$
$$I = 8.9 \text{ A}$$

The preceding example is an indication of why a high power factor is more desirable than a low one. A low power factor is a disadvantage in power circuits and should be corrected as soon as it is found. Because the properties of inductance and capacitance have opposite effects on an ac electric circuit, when circuits contain a large number of electric motors (a highly inductive circuit), installing large capacitors across the line will increase the power factor.

In most cases, electrical machinery is rated in VA (volt-amperes) rather than watts to help in determining the power factor. The use of a wattmeter will show the actual amount of power which is being taken from the power lines. The wattmeter reading divided by the full load VA for an electric motor indicates the power factor.

Chap. 7   Fundamentals of Electronic Control Systems

Semiconductors

Insulators offer a high resistance to the flow of electricity through an electric circuit. A conductor offers a small amount of resistance to the flow of current. Thus, the name semiconductor indicates a medium amount of resistance to the flow of current.

The semiconductors which are used in the making of diodes and transistors are basically made from germanium or silicon crystals containing small amounts of impurities which are added with great care. When the impurities arsenic or antimony are added, an *N*-type semiconductor material is the result. *N*-type semiconductors have an excess amount of electrons. When the impurities galium or indium are added, *P*-type semiconductor material is the result. *P*-type semiconductors have a lack of electrons.

Rectifiers (Diodes)

When the *N*-type and the *P*-type materials are joined together, a rectifier is formed. This rectifier is also sometimes referred to as a diode. The *PN* junction (diode) acts as a one-way valve to the flow of electrical current. There is a low-resistance direction through the junction known as a forward. The current which flows in the direction of the low resistance is called the forward bias. See Figure 7-12. Any current which flows in the direction of the high resistance is called a reverse bias. See Figure 7-13.

In practice diodes are used as rectifiers which convert ac to dc and for isolating one electrical circuit from another. See Figure 7-14. The output of the

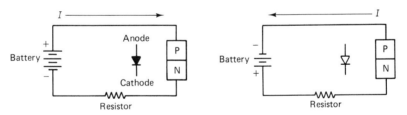

Figure 7-12 Forward bias diode.   **Figure 7-13** Reverse bias diode.

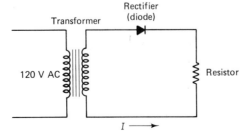

Figure 7-14 Simple rectifier circuit.

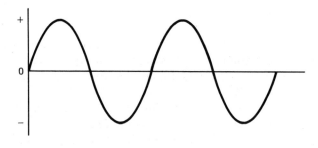

Figure 7-15 Transformer output.

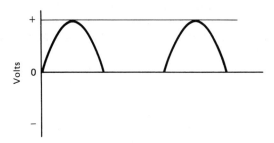

Figure 7-16 Voltage across a resistor.

transformer used in this figure is an ac voltage, as shown in Figure 7-15, but because the rectifier blocks current flow when the transformer output is negative, only pulsating dc voltage can be measured across the resistor. See Figure 7-16.

Zener Diodes

The zener diode, when compared to pneumatic equipment, is the equivalent to a nonadjustable pressure reducing valve. It is used in circuits where only dc voltage is required for proper operation.

When a voltage of one polarity is applied to a rectifier, the flow of current is blocked. If the voltage is raised high enough, however, the rectifier breaks down and allows the current to flow through. A normal type of rectifier would be ruined by this breakdown, but a zener diode is especially designed for operation in a breakdown situation.

Zener diodes are designed with a determined breakdown voltage. Let us consider a zener diode circuit with a breakdown voltage of 10 V. See Figure 7-17.

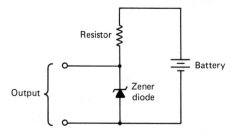

Figure 7-17 Zener diode circuit.

As long as the battery voltage is 10 V or more, the output across the diode will be 10 V. Any battery voltage above the 10 V is dropped across the diode. Thus, if the voltage is changed from 12 to 14 V, the voltage drop across the resistor changes from 2 to 4 V, while the input voltage remains at 10 V.

Silicon Controlled Rectifiers

Silicon controlled rectifiers (SCRs) are four-layered *PNPN* devices. The SCR devices are defined as high-speed semiconductor switches. They require only a short voltage phase for turn on as long as the current is flowing through it.

In operation, if we assume that the SCR is off (it has a very high resistance), it has no current flowing through it. See Figure 7-18. When the switch S_1 is closed for a time—just long enough to turn on the SCR (the SCR changes to a very low resistance)—an electrical current will flow through the resistor and the SCR. The SCR will remain in the on position until the switch S_2 is opened. When S_2 opens, it stops the flow of current through the resistor and the SCR. The SCR will then turn off. When the switch S_2 is closed again, the resistance of the SCR remains at a high value, and no current can flow through the resistor until the switch S_1 is closed again.

The anode is the positive terminal, and the gate is the terminal used to turn the SCR on. See Figure 7-19(a) and (b). Figure 7-19(a) shows the schematic diagram of the SCR, and Figure 7-19(b) shows the pictorial diagram of the SCR.

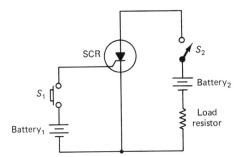

Figure 7-18 SCR operation.

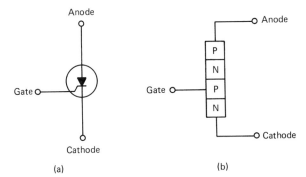

Figure 7-19 SCR schematic and pictorial diagrams.

Transistors

Transistors are devices which are used to amplify electronic signals. They are made of three layers of *P*- and *N*-type semiconductor materials which are arranged in either of two ways. See Figure 7-20.

The transistor is now rapdily replacing the vacuum tube in many of the electronic applications and is being used in many more different types of applications.

The transistor has many advantages over the vacuum tube type of system:

1. Low power consumption leading to economical operation.
2. Its ruggedness enables it to withstand mechanical shock and abuse better.
3. Its small size reduces the bulk and weight of electronic equipment.
4. No warm-up time is required, resulting in faster equipment response and operation.
5. Long life, which eliminates the frequent maintenance and replacement.

We can now trace the system operation of a simple transistor amplifier circuit. See Figure 7-21. In our example, we can consider that battery$_1$ and an adjustable resistor R_1 will determine the input current to the transistor. When R_1 has high resistance, the current flowing from the base to the emitter is very small. When the base-to-emitter current is small, the collector-to-emitter resistance appears to be very high, limiting the amount of current flowing from battery$_2$ and limiting the voltage drop across R_2.

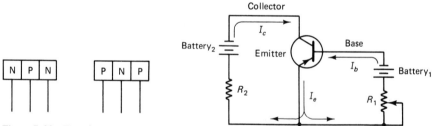

Figure 7-20 Transistor arrangement methods.

Figure 7-21 Simple transistor amplifier circuit.

As R_1 resistance is lowered, the amount of current flowing through the base-to-emitter junction is increased. As the base-to-emitter current is increased, the resistance of the transistor from the collector to the emitter is decreased. As more current flows from battery$_2$ through R_2, the voltage drop across R_2 is increased. A very small change in current flowing from battery$_1$ causes a large change in the amount of current flowing from battery$_2$. The ratio of the large change to the small change is defined as the gain of the transistor.

Bridge Theory

A bridge circuit is a system consisting of resistances and capacitive or inductive impedances and is generally used for making precise measurements. The most common is known as the Wheatstone bridge. This circuit consists of both variable and fixed resistances. This circuit is representative of a series-parallel circuit. See Figure 7-22. The branches of the circuit are called the legs.

If we were to apply 10 V of dc electricity to the bridge, one current would flow through resistors R_1 and R_2, and another would flow through resistors R_3 and R_4. See Figure 7-23. Because resistors R_1 and R_2 are fixed resistors having 1000 Ω of resistance each, the current flowing through them is constant, and each resistor will drop one-half of the battery voltage, or 5 V. As the resistance of R_4 varies, however, the voltage from the battery is divided between R_3 and R_4 in a different amount. Five volts is dropped across each resistor. The voltmeter senses the sum of the voltage drops across R_2 and R_3. Both of these voltage drops are 5 V; however, the R_2 voltage drop is a + to −5 V, and the R_3 voltage drop is a − to +5 V; thus they are opposite in their polarity and tend to cancel each other out. This is termed a balanced bridge. Usually this relationship is expressed as a ratio of $R_1/R_2 = R_3/R_4$. The actual values of the resistances are not important; what is important is that this ratio is maintained and that the bridge is always balanced.

If we change the variable resistor R_4 to 950 Ω and leave the remainder of the resistors at the same value, the circuit has become unbalanced. See Figure 7-24. We can use Ohm's law to find that the voltage drop across R_4 is 4.9 V; the remainder of the voltage, 5.1 V, is dropped across R_3. In our example, the voltmeter senses the algebraic sum of the voltage drop across R_2 and R_3, equal to −5.0 V and − to +5.1 V, indicating a total of −0.1 V.

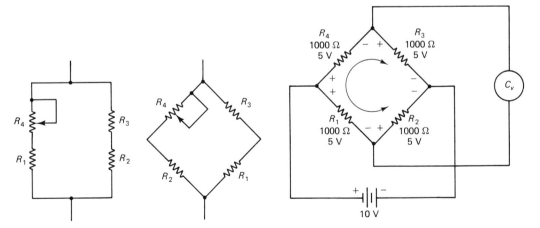

Figure 7-22 Wheatstone bridge circuit.

Figure 7-23 Wheatstone bridge circuit.

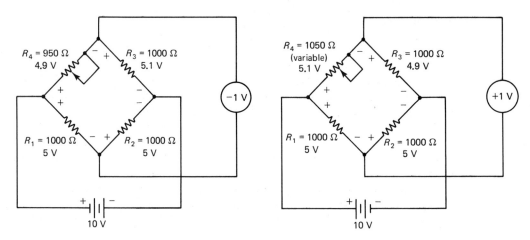

Figure 7-24 An unbalanced Wheatstone bridge circuit.

Figure 7-25 An unbalanced Wheatstone bridge circuit.

In another example, if we change the value of R_4 to 1050 Ω the voltage drop across R_3 is then 4.9 V. See Figure 7-25. The voltmeter senses the sum of + to −5 V and − to +4.9 V, or +0.1 V.

When the resistance of R_4 is changed the exact same amount above or below the balanced bridge resistance, the magnitude of the dc output as measured by the voltmeter is the same, but the polarity is reversed.

SENSORS

A sensor as used in a Cybertronic system is a resistance element whose value varies in accordance with any changes which occur in the variable that it is measuring. These resistance changes are converted into proportional amounts of voltage by a bridge circuit. This voltage is amplified and is used to position actuators that are used to regulate the controlled variable.

Temperature Elements

The temperature element which is used in the Cybertronic line of sensing devices is a nickel wire type of winding. This type of wire is extremely sensitive to changes in temperature. Its resistance is increased at the rate of approximately 3 Ω for every Fahrenheit degree increase in temperature. This type of action is called a positive temperature coefficient. The length and the type of wire give the winding a reference resistance of 1000 Ω at 70 °F. A drop in the temperature decreases the resistance, while a rise in the temperature increases the resistance.

Humidity Elements

Improvements in the design of humidity sensing elements and the materials which are used in their manufacture have reduced many of the limitations that humidity sensors experienced in the past. The sensors used with humidity controls in electronic control systems are a cellulose acetate butyrate (CAB) element. This type of resistance element is an improvement over the other resistance-type elements from the standpoint of contamination resistance, stability, and durability. The CAB element that is used in humidity controls is a multilayered humdity-sensitive polymeric film, made up of an electrically conductive core and insulating outer layers which are partially hydrolyzed. The element has a nominal resistance of 2500 Ω and a sensitivity of 2 Ω for each percent of relative humidity at 50% RH. The humidity sensing range is rated at 0–100% RH.

The CAB element consists of a conductive humidity-sensitive film, mounting components, and a protective cover. See Figure 7-26. The principal component of this humidity sensor is the film. See Figure 7-27. This film has five layers of cellulose acetate butyrate in the form of a ribbonlike strip. The CAB material is used because of its high sensitivity to humidity, good chemical and mechanical stability, and excellent film forming characteristics. See Figure 7-28.

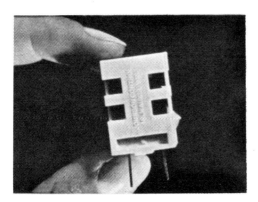

Figure 7-26 The CAB element. (*Courtesy of Robertshaw Controls, Uni-Line Division.*)

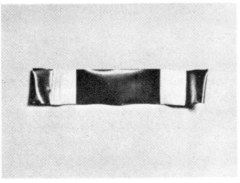

Figure 7-27 CAB sensing element. (*Courtesy of Robertshaw Controls, Uni-Line Division.*)

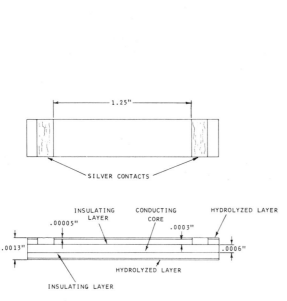

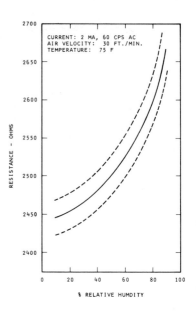

Figure 7-28 CAB sensing element construction. (*Courtesy of Robertshaw Controls, Uni-Line Division.*)

Figure 7-29 Comparison of humidity and resistance of CAB elements. (*Courtesy of Robertshaw Controls, Uni-Line Division.*)

The CAB resistance element is a carbon element having resistance/humidity tolerances. See Figure 7-29. With an increase in the relative humidity, water is absorbed by the CAB and causes it to swell. This swelling of the polymer matrix causes the suspended carbon particles to move farther away from each other, which results in a higher, or increased, element resistance. When the relative humidity decreases, water is given up by the CAB. The contraction of the polymer causes the carbon particles to come closer together and make the element more conductive, or less resistive.

CONTROLLERS

The sensing bridge of the controller circuit is the section that contains the temperature-sensitive element, or elements, and the potentiometer for establishing the *set point* of the control system. The control bridges are energized with a dc voltage. This permits longer runs of wires in the control circuit without the need for compensating wires or any other capacitive compensating schemes.

Both room- and remote-type sensing element controllers produce a proportional 0–16 V dc output signal in response to a measured temperature change. This type of controller can be wired to provide either direct or reverse action. The direct-acting type of operation provides an increasing output signal in response

Chap. 7   Fundamentals of Electronic Control Systems   113

to an increase in temperature. The reverse-acting type of operation provides an increasing output signal in response to a decrease in temperature.

Single-Element Controllers

There are three basic parts to electronic controllers:

1. Bridge
2. Amplifier
3. Output circuit

The bridge theory was discussed earlier in this chapter. The two legs of the bridge are variable resistances, that is, the sensor and the set point potentiometer. See Figure 7-30. In operation, if there is a change in the temperature, or if the set point is changed, the bridge is then in an unbalanced condition, and a corresponding output signal results. However, the output signal lacks the power required to position the actuators. This signal is therefore amplified.

The controllers use direct-coupled dc differential amplifiers to cause an increase in the millivolt signal from the bridge up to the required 0-16 V level for the actuators. There are two types of amplifiers used, one for direct-acting signals and one for reverse-acting signals. Each of these amplifiers has two stages of amplification. See Figure 7-31.

The differential transistor circuits provide a high gain and a good temperature stability. See Figure 7-32(a) and (b). Figure 7-32 shows a comparison of a single

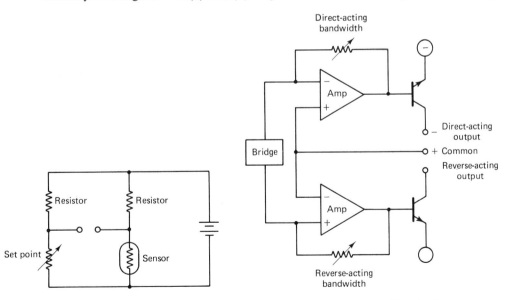

Figure 7-30 Bridge circuit with a sensor.

Figure 7-31 Amplifier circuit.

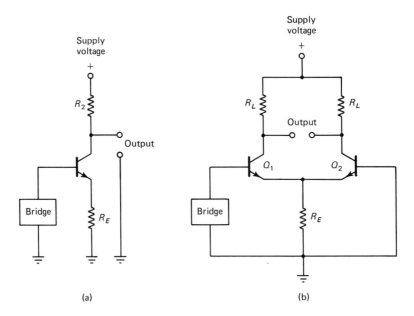

Figure 7-32 Differential transistor circuit.

transistor amplifier stage with a differential amplifier. Because transistors are temperature sensitive, the current that they allow to pass depends on the voltage which is present at the transistor and its ambient temperature. With an increase in the ambient temperature the current which is flowing through the transistor will increase, and the output voltage will decrease. See Figure 7-32(a). The emitter resistor R_E will reduce this temperature effect and also reduce the available voltage gain in the control circuit because the signal voltage across the resistor amounts to a negative feedback voltage; thus, it causes a decrease in the voltage difference which was originally produced by the change in temperature at the sensing element.

It is more desirable for the output voltage of the controller to correspond only to the temperature at the sensing elements and not to the ambient temperature of the amplifier, and so the circuit shown in Figure 7-32(b) is used. At this time, any ambient temperature changes will affect both of the transistors at the same time. The output that is used is the difference in the output levels of each transistor, and the effects of the temperature changes are cancelled out. The voltage gain shown in Figure 7-32(b) is a great deal higher than that shown in Figure 7-32(a). This occurs because the current variations in the two transistors which are produced by the bridge signal are equal but opposite in polarity. Any increase in current flow through Q_1 is accompanied by a decrease in current through Q_2. The sum of these currents through R_E is constant, and no signal voltage is shown at the emitters to cause a negative feedback, as shown in Figure 7-32(a).

Chap. 7   Fundamentals of Electronic Control Systems   115

There are three connections at the output circuit of the controller:

1. Common +, either solid red wire or terminal 15
2. Direct acting −, either solid blue wire or terminal 14
3. Reverse acting −, either white/blue wire or terminal 16

A load, such as an actuator, which is equivalent to 1000-Ω resistance, can be connected to either set of wires or to the terminals, depending on the type of action required of the controller.

The amplifier for the controller and the output circuits is also designed to provide the desired sequential operation of the two actuators. This action is obtained by connecting an actuator to each of the outputs, that is, to the direct- as well as to the reverse-acting outputs. The result is sequentially varying dc signals responding to a temperature change at the controller sensing element. See Figure 7-33. When this sequential operation is used, the controller is calibrated so that the set point and the sensing element will provide the necessary bridge circuit with a balanced condition at the set point, and therefore both the direct- and the reverse-acting outputs equal zero.

When the sensed temperature drops significantly below the set point, a 16-V dc output signal is present at the reverse-acting side and a 0-V dc signal is present at the direct-acting side. With an increase in temperature the reverse-acting signal decreases. When the temperature reaches the set point, both of the outputs are 0 V dc or at what is considered the *null* point. As the temperature continues to increase, the direct-acting signal is increased from 0 to 16 V dc. When the temperature has reached a point where operation is on the reverse-acting side of the null point, only the actuator that is connected to that side will be operating. Likewise, when the temperature is above the set point, the operation is on the direct-acting side of the null point, and only that actuator is operating. Thus, the actuators operate in sequence and not simultaneously.

The bandwidths which are used in these controllers are adjusted separately for direct- and reverse-acting signals. This type of adjustment permits optimum settings for both the heating and the cooling systems. See Figure 7-33.

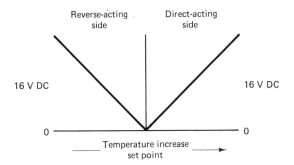

Figure 7-33 Varying DC signals of an amplifier.

The bandwidth adjustment of an electronic controller may be defined as the "number of desired degree changes at the element needed to cause a full 0-16 V dc change in the output signal." When sequential operation is used, the total amount of temperature change at the sensing element which caused the outputs of both of the sides of the null point to vary must be considered in the evaluation of the system control.

While there are two bandwidths, we must consider the temperature change from the set point to the point where the full 16-V output on each side of the null point should occur individually.

Dual-Element Controllers

The dual-element type of controllers operate the same as the single-element controllers except that there are two bridges rather than one included in the controller. Two bridge type of controllers are especially useful in applications where the temperature effects of one control element are to be used to readjust the set point of another element so that a greater accuracy in comfort control can be attained for the occupants. See Figure 7-34.

The bridge output of a dual-bridge arrangement is proportional to the algebraic sum of the temperature effects on both of the elements. This algebraic sum is expressed in terms of percent *authority*. For example, an authority of 100% simply means that a temperature change on the auxiliary element has the same effect as a temperature change on the main element, with the exception that the temperature change at each of the elements is in the opposite direction. This is referred to as a reverse adjustment.

The task of determining the main and the auxiliary assignments is dependent on the measured temperature span at each of the elements. The main element is always the element which has the least measured temperature change of the two elements, while the auxiliary element is always the element which has the greatest measured temperature change. This type of arrangement is essential because the authority settings are always between 0 and 100%.

A typical system would be one that might have a ratio of main to auxiliary sensor effects of 20 to 1, corresponding to a 5% authority setting. This indicates

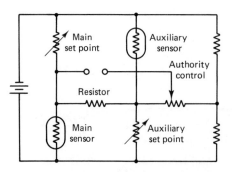

Figure 7-34 Dual bridge arrangement.

that a 20 °F change in the temperature at the auxiliary element will produce a bridge output which is equal to that of a 1 °F change in temperature at the main element. When a 2 to 1 ratio is used, the authority is 50%; that is, a 2 °F change in temperature at the auxiliary element will have the same effect as a 1 °F change at the main element.

The dual-element type of controllers differ from the single-element type only in regard to the bridge configuration, because there is an interacting effect within the bridge circuitry which is caused by the two elements and the authority setting. The amplifier circuitry and the output circuitry cause the signals on both sides of the null point to be identical to those which are encountered with single-element controllers.

ACTUATORS

An actuator may be defined as a "device which moves or stops the operation of the air conditioning equipment in response to changes in signals from the controller." The two types of actuators that we will discuss here are the electrohydraulic actuator and the thermal actuator.

Electrohydraulic Actuators

Cybertronic types of actuators do the work in an electronic control system. They take the control signal from the controller and translate it into a mechanical movement to operate or position the valves or the dampers.

The electrohydraulic actuators are named so because they convert an electrical signal into a fluid-type movement and its force. Damper operators which are equipped with a linkage for connection to dampers and valve actuators which have a yoke and a linkage to aid in mounting them on the valve body are available.

Operation: In the operation of electrohydraulic actuators, there are two voltages which are applied to the actuator, a 0–16 V dc control signal which regulates or controls the servo valve and 24- or 120- V ac control signal, depending on the type of unit, which operates the oil pump. The oil pump moves the oil from the upper chamber to the lower chamber. The servo valve controls the pressure at the diaphragm by varying the return flow of the oil from the lower to the upper chamber. See Figure 7–35.

In applications where there is no dc voltage applied to the servo valve, the flapper is pushed off the servo port by the hydraulic pressure developed by the oil pump. The open servo port allows the oil pump to move all the oil through the lower chamber and back into the upper chamber.

As the voltage increases on the servo, a magnetic force is caused which holds the flapper down over the servo port. The oil pump continues to pump the oil into the lower chamber, but the return oil flow to the upper chamber is stopped by the blocked servo port. Oil pressure is built up in the lower chamber until

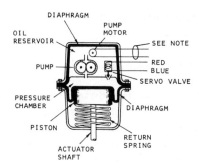

Figure 7-35 Cut-away of an electro-hydraulic actuator (*Courtesy of Robertshaw Controls, Uni-Line Division.*)

the magnetic force on the flapper valve is overcome and the flapper is pushed from the port. This then allows the oil to flow through the pump and the servo valve, while at the same time maintaining an oil pressure in the lower chamber.

Each time there is an increase in the dc voltage, there is a resultant increase in the hydraulic pressure in the lower chamber. This increase in pressure begins to overcome the opposing pressure from the return spring and forces the actuator shaft out. Each time there is an additional increase in the dc voltage, there is an increased extension of the actuator shaft.

The servo valve represents an electrical load of 1000-Ω resistance which is required by the controller to cause a variation of the 0–16 V dc output signal. It is possible to connect two actuators in parallel across the output terminals of an electronic controller. This procedure will provide only 500-Ω resistance, which the controller can also handle properly.

Thermal Actuators

A more proper name for the thermal actuator is electrothermal, because the actuator takes 0–16 V dc signal and converts that signal into heat. The thermal damper actuator has a linkage for connection to a damper. The thermal valve actuator is thus directly connected to a valve body.

Operation: During the installation, a small electrical circuit is encapsulated in the electrical cable about 12 in. from the thermal unit. The 0–16 V dc signal from the controller and the 24-V ac supply voltage are fed into the control circuit. The circuit allows the 0–16 V dc signal to control the amount of electrical current flowing from the 24-V supply to the actuator. See Figure 7-36.

The controlled current from the 24-V source heats up a small heater which is embedded in wax inside the actuator. When the wax is heated to approximately 180°F, it is changed from a solid to a liquid. The wax expands during this change of state. At this point the motion of the device is controlled.

This expansion of the wax forces the power element shaft to move the piston, which in turn compresses the return spring and moves the actuator shaft. When the power element shaft has traveled its full stroke, a limit switch opens, which

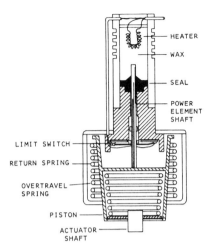

Figure 7-36 Cut-away of a thermal actuator. (*Courtesy of Robertshaw Controls, Uni-Line Division.*)

interrupts the flow of electricity to the heater. As the wax begins to cool, it also contracts, and the power element shaft is forced back into the valve body by the return spring. This action allows the limit switch to close, and the sequence is repeated when the control signal has reached a high enough voltage to hold the actuator at its fully extended position.

AUXILIARY DEVICES

Auxiliary devices such as the low- and high-signal selectors accept several types of control signals, compare them, and pass either the highest or the lowest signal on to the controlled device. As an example, high-signal selectors can be used on multizone applications to control the cooling coil. The zone which requires the most cooling transmits the highest control signal, which will be passed along by the high-signal selector to energize the cooling apparatus.

In some installations, minimum positioning networks are used to make certain that the outdoor air dampers are properly positioned to allow a minimum amount of outside air in for ventilation purposes regardless of the controller demand.

Reversing networks are used to change the action of a controller output signal from direct acting to reverse acting or from reverse acting to direct acting.

Sequencing networks are used to amplify only a selected portion of an input signal from a given controller. These devices are especially useful in applications where two actuators operate in sequence.

Two-position power supplies to auxiliary devices allow a two-position override of a proportional control system.

Another type of auxiliary device is the unison amplifier which allows a controller to operate up to eight actuators. This is useful because a controller alone can operate only two actuators.

REVIEW QUESTIONS

1. How does electronic control equipment differ from electric control equipment?
2. What type of circuit requires that the same current flow through all the circuit resistances?
3. In what type of electrical circuit is the voltage drop across each of the branch lines always the same?
4. How is the frequency of ac current or voltage determined?
5. Is the heating effect of electricity dependent on the direction of the current?
6. How are most of the ac ammeters and voltmeters calibrated?
7. What is electrical force called?
8. How is electrical flow related to the amount of work done?
9. What property in an electric circuit, or a circuit component, opposes any change in the current flowing through it?
10. How does capacitance affect an electric circuit?
11. How are capacitors rated in air conditioning units?
12. What is the impedance of an electric circuit?
13. What does the name semiconductor indicate?
14. How is a rectifier formed?
15. How are rectifiers used in electronic control circuits?
16. How does a zener diode compare to pneumatic equipment?
17. How are bridge circuits normally used?
18. What type of elements are used in electronic humidity control systems?
19. Name the three basic parts of an electronic single-element controller.
20. What is the purpose of direct-coupled dc differential amplifiers?
21. Define the bandwidth adjustment of an electronic controller.
22. What is the difference between single-element and double-element electronic controllers?
23. What is the purpose of electrohydraulic actuators?
24. What are the two voltages which are applied to electrohydraulic actuators?
25. How do thermal actuators work?

8

Basic Electric Control Circuits

The use of electric energy in transmitting the controller's measurement of some change in the controlled variable to the controlled parts of the system is common practice. It is also used to translate that measurement into useful work at the final control element. Because of these things, electricity has the following advantages:

1. Electricity readily amplifies the relatively weak signal which is received from the sensing element in the controller, thus making it possible to properly control systems which would ordinarily be difficult to control.
2. Electricity readily allows the system to be controlled from some remote point. Thus, the controller can be located at some distance from the final control element.
3. The installation of electric wiring is usually very simple.
4. The signal which is received from the controller element can be directly applied to produce one or several different combinations or sequences in the electrical output. Therefore, one single actuator can be used for several different desired functions.
5. Electricity is available wherever power lines may be installed.

DEFINITIONS

The definitions which are listed below are applicable to electric control systems.

Line Voltage

Line voltage is the term that applies to the normal electrical service supplying voltage in the range of 120 or 240 V. The line voltage may be used to perform the desired work directly, or it may be connected to the primary winding of a transformer which will reduce the line voltage to low voltage for the control circuit. The terms *high voltage* and *line voltage* are usually used interchangeably. This sometimes leads to some confusion because the utility companies use the term high voltage when referring to 500 V or more in their system.

Low Voltage

The term low voltage, in automatic control terminology, refers to the wiring of the other electrical devices which use 25 V or less in operation. Most of the control manufacturers use the 25-V type of controls in their systems.

Potentiometer

A potentiometer is made from a number of turns of a highly resistance wire which is wrapped around some type of core and has connections provided for connecting to the system. See Figure 8-1. The center connection is a movable finger, or wiper blade, which moves over the entire length of the coil and makes a complete circuit wherever it touches the coil. The blade can be moved along the coil either manually or automatically.

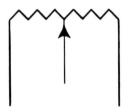 

Figure 8-1 Typical potentiometer symbols. (*Courtesy of Honeywell Inc.*)

Balancing Relay

A balancing relay is made up of a relay which has a pivoted armature which is allowed to swing between two electromagnetic coils. See Figure 8-2. As the electric current through the two coils changes, the magnetism of the coils also changes, and the armature will move toward the stronger of the two magnetic fields. In most cases the armature relay is equipped with two contacts which are adjusted so that certain circuits are completed at the proper positions of the armature.

In Contacts and Out Contacts

The relay contacts which make an electric circuit when the armature is pulled into the relay coil when it is energized are called *in* contacts. Those contacts which

Chap. 8 Basic Electric Control Circuits

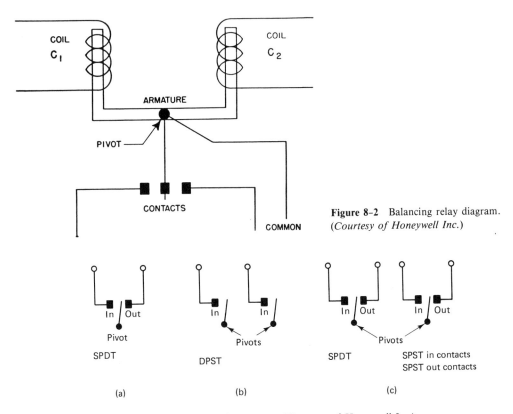

Figure 8-2 Balancing relay diagram. (*Courtesy of Honeywell Inc.*)

Figure 8-3 Relay contact. (*Courtesy of Honeywell Inc.*)

complete a circuit when the relay coil is de-energized are called *out* contacts. We will show the relay contacts in the out position unless they are otherwise designated. See Figure 8-3.

CLASSIFICATION OF ELECTRIC CONTROL CIRCUITS

In present-day control systems, there are seven basic electric control circuits, each one having its own characteristics and functions. These different types of control systems are identified by Minneapolis-Honeywell as Series numbers 10, 20, 40, 60, 70, 80, and 90.

Table 8-1 is a classification and description of the different control circuits, as listed by Minneapolis-Honeywell.

The construction of each of the individual controls meets the specific requirements for use in the basic circuit for which the control was designed. Generally, controls of a specific series would be used in a control circuit with the same designated number. *Example:* A Series 10 control would be used in a Series 10 circuit. There are, however, many types of applications for limit controls of a

TABLE 8-1 CONTROL CIRCUIT CLASSIFICATION (*Courtesy of Honeywell Inc.*)

Series	Controller	Signal Circuit	Actuator	Control Mode
10	Two sets of contacts make in sequence to start, break in reverse sequence to stop	Three-wire, low voltage	Any Series 10 motor or solenoid	Two-position
20	Makes one circuit to start, breaks it, and makes a second circuit to stop	Three-wire, low voltage	Any Series 20 device	Two-position
40	Makes circuit when switch is closed, breaks it when switch is open	Two-wire, line voltage	Any Series 40 device	Two-position
60 two-position	Line-voltage equivalent of Series 20	Three-wire, line voltage	Series 60 two-position control motor	Two-position
60 floating	Line- or low-voltage floating control	Three-wire	Series 60 floating motor	Single-speed floating
80	Low-voltage equivalent of Series 40	Two-wire, low voltage	Any Series 80 motor or solenoid	Two-position
90	Modulating action	Three-wire, low voltage	Any Series 90 motor	Proportional

series different from the control circuit in which they are installed. There will be many variations shown later in this chapter.

The preceding information has made reference to and has listed and described the basic control circuits. In operation, these basic arrangements are sometimes expanded to provide additional features such as

1. High-limit protection
2. Low-limit protection
3. Compensated control (mechanical or electrical)
4. Positive cycling sequence

The following information will introduce some of the more common applications of these controls. When it is desirable to use additional controls to achieve the functions which were listed above, it should be noted that the basic control circuit is not changed.

To present more understandable information, the simplest control circuits will be introduced first.

Series 40 Control Application

The Series 40 control circuit uses line voltage which is switched directly by the single-pole single-throw (SPST) switching action of a Series 40 controller. The Series 40 control circuit is operated by a two-position type of controller and requires two wires. It is useful for controlling lights, fans, electric motors, and other standard line-voltage-type equipment in addition to the Series 40 control motors and relays which are especially designed for use on Series 40 circuits. For a Series 40 circuit to provide fail-safe operation, it must be used with the proper type of controlled equipment.

In operation, the equipment being controlled is energized when the controller switch is closed, and it is de-energized when the switch is opened. This arrangement is the most simple in principle.

In most control circuits, the Series 40 controller makes and breaks the controlled load directly. It is possible, however, for the loads to exceed the current rating of the controller. When this is possible, a simple intermediate relay is installed between the load and the controller.

A relay may also be used when the load requires more than one switching action. In these instances, the Series 40 controller energizes and de-energizes the relay coil, which in turn opens and closes the desired number of contacts in the circuit.

The Series 40 types of controllers are designed to operate on line voltage and therefore should be installed and wired accordingly.

Series 40 Control Equipment

The different types of Series 40 equipment are listed and described below.

Controllers: The following is a listing of the different types of Series 40 controllers:

1. Room thermostats
2. Insertion thermostats
3. Pressure controllers
4. Humidity controllers

These controllers may be equipped with either open-type contacts or mercury switches which are incorporated in them to make and break an electric circuit. In most cases, the Series 40 controllers make use of the snap-acting-type switch.

Relays: The Series 40 relay is made up of a line voltage coil which operates an armature carrying one or more sets of in or out contacts that are capable of handling the circuit current load.

Motor units: A Series 40 motor unit may be adapted, through a linkage, to operate the desired valves or dampers. These motors are electrically driven through their complete stroke. The stroke may be either 60° or 160°. This action takes place when the controller contacts are closed. When the end of the power stroke is reached, the motor stops and is then held in this position by a holding coil as long as the controller keeps the circuit closed. When the circuit is broken by the controller or when a power failure occurs, either an internal or an external return spring returns the motor to the de-energized position. This action makes the Series 40 controller when used on motor applications fail-safe.

Solenoid valves: Series 40 valves are designed to, and usually do, go closed when they are de-energized, thus providing a fail-safe operation.

Series 40 Control Operation

A simple Series 40 control circuit consists of a Series 40 controller and a Series 40 motor. See Figure 8-4. In a Series 40 control circuit, the internal circuit of the controlled device is not considered except as a means of making the electrical connections.

When the controller contacts are open (See Figure 8-4), the current is inter-

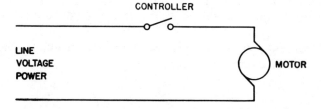

Figure 8-4 Series 40 control circuit. (*Courtesy of Honeywell Inc.*)

Chap. 8   Basic Electric Control Circuits   127

rupted to the valve, motor, relay, and other controlled devices. When the controller contacts are closed, the electrical circuit is completed to the controlled devices.

Series 40 Control Combinations

Since the Series 40 control circuits are so simple, the control combinations are also extremely simple.

Unit heater control: Control of a unit heater is necessary to provide some means of keeping the fan from running when the burner is not on to prevent the circulation of cold air into the space. See Figure 8-5. This is a typical Series 40 unit heater wiring diagram. Even though it may have the same appearance as the one presented in Figure 8-6, this diagram does not accomplish the same purpose.

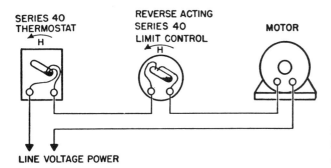

Figure 8-5 Series 40 unit heater control diagram. (*Courtesy of Honeywell Inc.*)

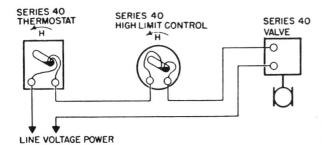

Figure 8-6 Series 40 high-limit control diagram. (*Courtesy of Honeywell Inc.*)

High-limit control: The Series 40 high-limit control is wired in series with the thermostat and the valve so that it will either make or break the electrical circuit to the valve on a rise in the temperature inside the unit heater. See Figure 8-6. In this circuit, both the thermostat and the high limit must be snap acting to prevent a high sensitivity or short cycling of the unit.

The Series 40 limit control is usually a reverse-acting pressure control installed with its sensing element located in the steam supply line to the unit heater. Since it is connected in series with the thermostat and the motor, it will not let

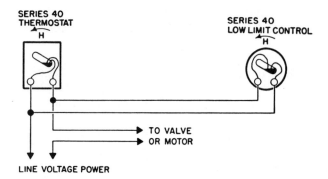

Figure 8-7 Series 40 low-limit control system diagram. (*Courtesy of Honeywell Inc.*)

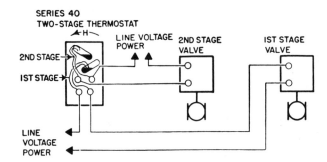

Figure 8-8 Series 40 two-stage control system. (*Courtesy of Honeywell Inc.*)

the motor run until the steam pressure has risen to a predetermined point.

Low-limit control: The low-limit controls are connected in parallel with the primary controller. See Figure 8-7. The low limit-control can make the electrical circuit through the steam valve even though the primary controller has interrupted the circuit.

Two-stage control: The Series 40 two-stage control is usually used in air conditioning units to control two stages of the refrigeration unit in sequence. See Figure 8-8. When a rise in temperature is sensed, the lower mercury bulb tips to the left and interrupts the first-stage cooling valve. If the temperature should drop more, the upper mercury bulb tips to the left and interrupts the circuit to the second-stage cooling valve.

Series 80 Control Application

Series 80 low-voltage control systems are the equivalent of the Series 40 circuits. They are suited for applications which require a low-voltage two-position type of control for a two-wire circuit. They are single-pole single-throw switches.

The Series 80 control circuits have two advantages over the Series 40 control circuits:

1. The electrical contacts of a Series 80 controller are required to carry a small amount of current; therefore, they and the controller can be made of a smaller

Chap. 8 Basic Electric Control Circuits

mass than the Series 40 units. Because of this, the controller has less lag and also can be made to respond to a narrower differential controller.

2. The required low-voltage wire, when not installed in a conduit or an armor cable, has a lower cost than the high-voltage wiring.

The most popular Series 80 actuators are solenoid valves, water valves, and damper motors which are designed to operate in these circuits.

Series 80 Control Equipment

The following is a list of the Series 80 equipment and its description:

1. Room thermostats
2. Insertion thermostats

Series 80 controllers are not designed to switch line voltage directly. There must be a relay installed between the controller and the line-voltage load. Series 40 controllers can, however, be used with Series 80 actuators.

Series 80 controllers must have snap-acting contacts or a mercury switch. Slow-moving-type contacts are not suitable for use in these controllers.

Relays and motors: The relays and motors used in Series 80 circuits are similar to those which are used in the Series 40 circuits with the exception that they are designed for use on low-voltage electricity.

Valves: Generally, the valves used in Series 80 circuits are similar to those which are used in the Series 40 circuits except that the Series 80 controls are for low voltage.

There is available a self-powered gas valve which operates on approximately $\frac{1}{2}$ V furnished by a thermopile. This $\frac{1}{2}$ V is lower than the 24 V supplied by a transformer.

Series 80 Control Operation

The diagrams shown in Figure 8-9(a) and (b) are two simple Series 80 circuits. It should be noted that they are the same diagrams as those for the Series 40 circuits previously shown except that there is a transformer included in these diagrams.

Heat anticipation: A *resistance heater* is included in most of the Series 80 thermostats. See Figure 8-9(a). This resistor provides enough artificial heat during the on cycle to cause the bimetal to open the mercury switch in response to a smaller rise in the temperature of the space than would ordinarily be required to open it. This action tends to flatten out the override spots which are normally produced by a measurement lag and the flywheel effect of the heat storage capacity of the building.

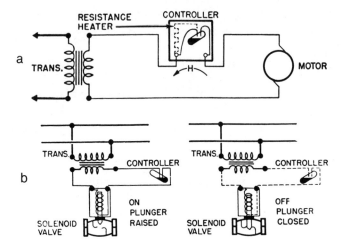

Figure 8-9 (a) Series 80 motor application. (b) Series 80 valve application. (*Courtesy of Honeywell Inc.*)

Series 80 Control Combinations

The possible Series 80 control combinations are similar to those of the Series 40 type. There are a few exceptions, however, as follows:

1. The Series 80 circuits require a transformer.
2. The Series 80 equipment can be controlled by a Series 40 or a Series 80 controller, but the Series 40 equipment must be controlled by a Series 40 controller.

Solenoid valve control: The Series 80 solenoid valve and high-limit control are shown in Figure 8-10. In this diagram, the two-wire controller and the high-limit control operate a Series 80 valve. Either a Series 40 or a Series 80 controller and high-limit type of control can be used. The solenoid valve is a Series 80, which requires a transformer. The high-limit control interrupts the electrical circuit on an increase in the temperature.

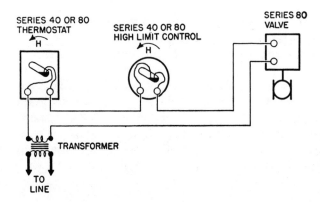

Figure 8-10 High-limit control of series 80 solenoid valve. (*Courtesy of Honeywell Inc.*)

Chap. 8   Basic Electric Control Circuits   131

This combination can also be used to control other types of Series 80 equipment.

Low-limit control: This is the same as that used for the Series 40 with the exception that a transformer and a Series 80 actuator are required.

Series 10 Control Application

The circuits which are designed for Series 10 controls are for two-position operation of the following controls:

1. Valves
2. Relays

Series 10 equipment is designed to return to its de-energized position in the event of a power failure, thus making them fail-safe in operation.

A basic Series 10 control circuit is made up of a controller to which is connected a relay, a motorized valve, or a solenoid valve. Should limit controls be required for a particular application, they can be added to the basic Series 10 control circuit with little difficulty.

Series 10 Control Equipment

The following equipment is used in the Series 10 control circuits.

Controllers: The different Series 10 types of controllers are as follows:

1. Room thermostats
2. Insertion thermostats
3. Pressure controls

The primary element which is used in the Series 10 controllers operates with two bimetal blades. Each of these blades has its own set of contacts. The bimetal blades engage with their own specific set of contacts in a sequence as the controlled condition reaches the two predetermined points. First a starting circuit is completed, and then a holding circuit is completed. The blades disengage their contacts in the reverse sequence. As the controlled condition becomes satisfied, first the starting circuit is broken, and then the holding circuit is broken.

Relays: The Series 10 relays are operated by the electromagnetism principle. The relay is made up of the following components:

1. A low-voltage starting circuit
2. A low-voltage holding circuit
3. A built-in transformer
4. A line-voltage switching circuit

The built-in transformer supplies the electric current for the operation of the starting and the holding circuits. The current is supplied at a continuous rate on demand by the circuit unless there is a manual shutoff included in the circuit.

When both of the blades in the controller are engaged with their contacts, the relay starting circuit is energized. This circuit causes the armature to be pulled in against the relay coil and at the same time closes both sets of the contacts. One set of the contacts energizes the holding circuit, while the other set energizes the load circuit.

Valves: The normal operation of a Series 10 valve is identical to that of the Series 10 relay with the exception that the coil of the valve operates a plunger or a motor armature rather than a set of line-voltage contacts located in the relay. The low-voltage contacts which engage the holding circuit are also used in the valve assembly. The load contacts which are used in the relay are not used in the valve assembly because the valve itself is acting as the final control element.

Series 10 Control Operation

A complete basic Series 10 control circuit includes the following equipment:

1. Room thermostat
2. Relay

In Figure 8-11 the thermostat is satisfied, and both sets of movable contacts are positioned away from their stationary contacts. The bimetal blades are arranged so that the flexible bimetal will make contact with the white (W) terminal and the rigid bimetal will make contact with the blue (B) terminal.

Both of the relay contacts (1 and 2) are open. See Figure 8-11. For simplicity, they are shown separately in the diagram, but in actual practice they are mechanically connected so that they will open and close at the same time.

Figure 8-11 shows only the line-voltage circuit. Figures 8-12 through 8-17 show the low-voltage circuit as it takes place in both the relay and the valve.

When the thermostat is positioned above the set point (see Figure 8-12),

1. Neither of the bimetal blades is touching its contact point.
2. All of the control circuits are open. No current is flowing.

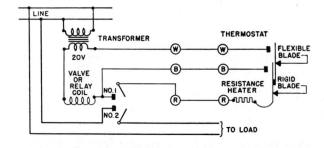

Figure 8-11 A complete basic series 10 control circuit. (*Courtesy of Honeywell Inc.*)

Chap. 8   Basic Electric Control Circuits

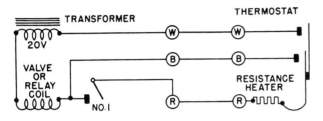

Figure 8-12 Thermostat is satisfied. (*Courtesy of Honeywell Inc.*)

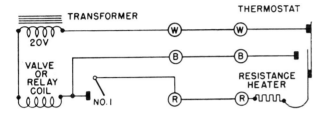

Figure 8-13 Slight drop in room temperature. (*Courtesy of Honeywell Inc.*)

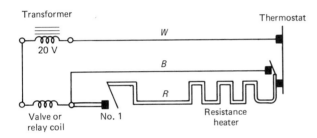

Figure 8-14 A further drop in room temperature. (*Courtesy of Honeywell Inc.*)

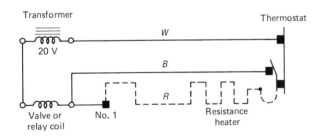

Figure 8-15 Holding circuit established. (*Courtesy of Honeywell Inc.*)

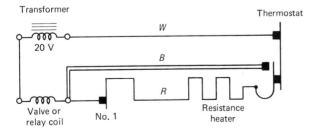

Figure 8-16 A slight rise in temperature. (*Courtesy of Honeywell Inc.*)

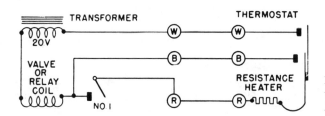

Figure 8-17 A further rise in temperature. (*Courtesy of Honeywell Inc.*)

3. The relay or the valve coil is de-energized (whichever component is used). The contacts 1 and 2 are open (2 is not shown).

When there is a slight drop in the room temperature, the following things occur (see Figure 8-13):

1. The bimetal bends toward the stationary contacts and the flexible blade comes into contact with the *W* contact.
2. All of the other circuits remain open.

When the room temperature continues to drop, the following procedures take place. See Figure 8-14. The following line markings are used in the next three diagrams.

1. A solid black line indicates that the maximum current is flowing through that part of the control circuit.
2. A broken black line is an indication that a negligible amount of current is flowing through that part of the control circuit.
3. A double black line is an indication that part of the control circuit is inactive.

The following is a description of the operation of the thermostat when this condition exists:

1. The thermostat bimetal bends a little further, and the rigid contact blade comes into contact with terminal *B*.
2. The control starting circuit is now established.
3. The valve coil or the relay is now energized.

When the holding circuit is established, the following is a description of the operation. See Figure 8-15. The energized coil pulls in the armature an instant later, contact 1 is closed, and the holding circuit is then completed. Because of the heater element resistance in the holding circuit, most of the current continues to flow through the starting circuit. The current, which is indicated by the broken line, is not strong enough to increase the temperature of the heater element to any great extent. Contact 2 is also closed at the same time that 1 is closed. See Figure 8-11.

When a slight rise in the room temperature is sensed, the following things occur (see Figure 8-16):

1. The rigid thermostat bimetal blade breaks contact with the *B* terminal, thus breaking the starting circuit.
2. The full amount of current continues to flow through the resistance heater in the holding circuit, and the bimetal is heated.
3. The relay coil continues to hold in contacts 1 and 2.

As the room temperature continues to rise, the conditions exist as shown in Figure 8-17. The rise in room temperature is aided by the increase in temperature of the resistance heater along with the following actions:

1. The flexible blade breaks contact with the *W* contact.
2. The relay is de-energized and therefore drops out.
3. Both the holding and the load circuits are broken.
4. The heater element is also de-energized.

Series 10 Control Combinations

In Series 10 circuits, limit control may be accomplished in several different ways. In some applications, the limit control may consist of a mercury switch type of controller which handles line voltage, while in other applications it may be a Series 10 open-contact controller switching low voltage. See Figure 8-18. This figure is an example of how a Series 40 mercury switch low-limit control can be connected into a Series 10 control circuit.

High- and low-limit Series 10 control circuit: Series 10 valves and relays can be operated by more than one two-wire controller. The controllers, however, must be equipped with mercury switches or snap-acting contacts. A slow switching action is not suitable for this type of control circuit.

For a diagram representing the connection method of a room thermostat, a low-limit control, and a high-limit control, see Figure 8-19. Note that all of these controls are of the two-wire type and are used to operate a Series 10 relay.

Two controllers and one relay: A common application is one which uses

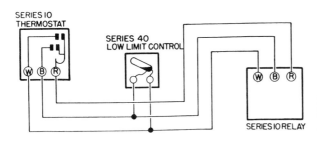

Figure 8-18 Series 40 low-limit control in a series 10 circuit. (*Courtesy of Honeywell Inc.*)

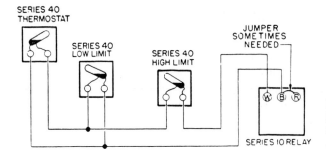

Figure 8-19 Series 40 thermostat and high and low limit controls controlling a series 10 relay. (*Courtesy of Honeywell Inc.*)

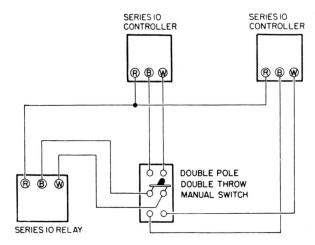

Figure 8-20 Two series 10 controllers controlling one series 10 relay. (*Courtesy of Honeywell Inc.*)

two Series 10 controllers and one Series 10 relay. See Figure 8-20. Only one controller has control over the relay at a time. There is a double-pole double-throw manual switch required to provide the proper changeover action because two wires must be broken for a Series 10 controller to be completely inoperative. A relay or a time clock equipped with a DPDT switching action can be used rather than the manual switch as shown.

Warning: Do not attempt to control more than one Series 10 device at any one single time from a single Series 10 controller. The circuit action between the two controlled devices will prevent dropping out of the relays.

Series 60 Floating Control Application

The action of Series 60 floating control circuits is different from those previously described, because they provide a floating action rather than the two-position operation. The Series 60 floating type of control is more commonly applied to

1. Motorized valves which are used on tank-level control systems
2. Motorized dampers which are used for static pressure regulation

Chap. 8   Basic Electric Control Circuits   137

3. Specialized pressure and temperature control systems

Series 60 floating-type control circuits are adaptable on either line voltage or low voltage. This factor is determined by the type of equipment which is selected. The basic temperature pattern is similar to that of the two-position control circuit with the exception that the motor is reversible and the limit switches are substituted for maintaining switches.

In operation, the floating-type control circuit has no fixed number of final control element positions. The controlled device, either the valve or the damper, can assume any position between the two extremes as long as the controlled variable remains within the values which correspond to the neutral zone of the controller. Also, if the controlled variable should drift outside of the neutral zone of the controller, the final control element will move toward the corrective position until the value of the controlled variable is moved back into the neutral zone of the controller or until the final control element has reached its extreme position on the controller.

This type of control system is best used where the process lag is very short and in applications where it is possible to use a controller having a very small amount of lag.

Some of the more common types of applications are in static pressure regulation using motorized dampers, in liquid-level control, and in the control of the suction pressure of several refrigeration compressors which are connected to a single evaporator.

It should be noted that the Series 60 floating-type controls are not generally applied to dampers in a thermal system because the process lag is far too great and they can be controlled with other types of control systems with much better results.

The Series 60 floating control system is a single-speed-type circuit, and thus the actuator moves toward its new position at a single predetermined rate. This rate must be timed to match the natural cycle of the controlled system. Should the actuator move too slowly, the control system would not be able to keep pace with the sudden changes. If the actuator should move too rapidly, two-position control operation would be the result. In most applications, the actuator should move just at the right speed to keep pace with the most rapid changes in the load that can possibly occur.

We can consider a damper which is installed in a fan heating system and which is equipped with a motorized damper motor to regulate the amount of supply air going into the fan intake. Control of the damper is accomplished by using a static pressure regulator measuring the air pressure in the distribution duct. With a change in the duct static pressure the static pressure controller senses this change and causes the damper to be positioned to cause the static pressure to come back to the value which is within the neutral zone of the controller.

In this situation, when a two-position type of control system is used, the damper would either be fully closed or fully open in response to the amount of static pressure inside the duct. The floating-type control system places no limitation on the number of damper positions, and therefore the damper may be positioned at any point in the travel path which will allow enough air to pass for the fan to maintain the desired duct static pressure. Should the amount of air passing through the damper be insufficient to maintain the desired duct static pressure, the controller will open the damper and allow the proper amount of air to pass. In this mode, the damper motor will continue to move the damper toward the open position until there is enough air passing to maintain the static pressure within the required limits of the controller. At this point, if the static pressure is held constant, the damper will hold this position. However, if the static pressure rises above the desired limits, the damper motor will move the damper to reduce the amount of air until the pressure is held within the desired limits.

In this manner the damper will move toward the open and the closed positions as long as the duct static pressure is either above or below the set limits of the controller.

This circuit has no holding or maintaining switch action, and therefore the damper is not required to move to its limit and can stop at any position that is required to maintain the duct air static pressure within the limits of the controller. The damper thus floats between the outer limits of its travel path as the controller maintains the desired static pressure between the prescribed limits.

In this system, an accurate pressure can be maintained with a minimum amount of offset.

Series 60 Floating Control Equipment

The controllers used in the Series 60 floating-type control circuits are arranged so that a single-pole double-throw switching action can be provided.

Since it is required that the controller have a slow action so that the switch blade will float between the two contacts, slow-acting controllers are also used. However, snap-acting controllers cannot be used in these circuits.

The line-voltage floating type of controller can be used on either line-voltage- or low-voltage-type motors. However, line-voltage-type motors require that line-voltage controllers be used.

Motor units: The Series 60 floating control motor units are made up of the following:

1. A reversible capacitor power unit in both the line- and the low-voltage units.
2. Limit switches are used to limit the rotation of the crank arm.
3. A gear train to transfer the power from the motor armature to the drive shaft.

A power unit wiring diagram will show the coils W_1 and W_2, the capacitor M, and the circuit connections. See Figure 8-21. The common, or red, terminal is connected to one side of the electrical power line. Motor operation can be

Chap. 8 Basic Electric Control Circuits

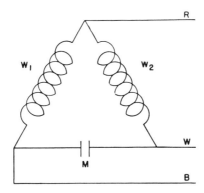

Figure 8-21 Internal circuit of the series 60 motor diagram. (*Courtesy of Honeywell Inc.*)

achieved when the other side of the power line is connected to either the W terminal or the B terminal. The direction of this rotation is dependent on which terminal the power line is connected to. Therefore, the direction of the motor can be changed by the switching action of the controller as it directs the electric power to either the W terminal or the B terminal when a corrective action is required at the final control element.

When the controller is in a position such that the W terminal is energized, the electric current will flow directly through the W_2 winding of the motor. At this time current is also flowing through the W_2 terminal, but it must first flow through the capacitor M. The electric current flowing through W_2 has a phase shift in relation to that flowing through W_1, and the motor is caused to turn in the corresponding direction. As the controller switch moves to a position to energize the B terminal, the phase of the electric current is shifted in the other direction, and the motor direction is reversed. The motor remains stationary when there is no current directed to either the W or the B terminal.

There are limit switches mounted in the motor to break the W and B circuits to stop the motor when it reaches its full travel limit. They also limit the rotation of the motor crank arm to 160° in this manner.

Series 60 Floating Control Operation

A complete Series 60 floating control circuit includes the following equipment (see Figure 8-22):

1. A static pressure controller
2. A Series 60 floating control motor

The following is the action on a drop in pressure (see Figure 8-23):

1. The controller blades contact the B terminal, making R to B.
2. The B terminal of the motor is energized.
3. The motor is caused to rotate in a direction that will cause the corrective action of the final control element.

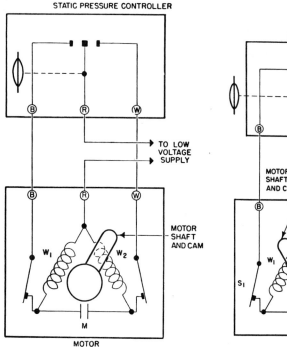

Figure 8-22 Diagram of a complete series 60 floating control circuit. (*Courtesy of Honeywell Inc.*)

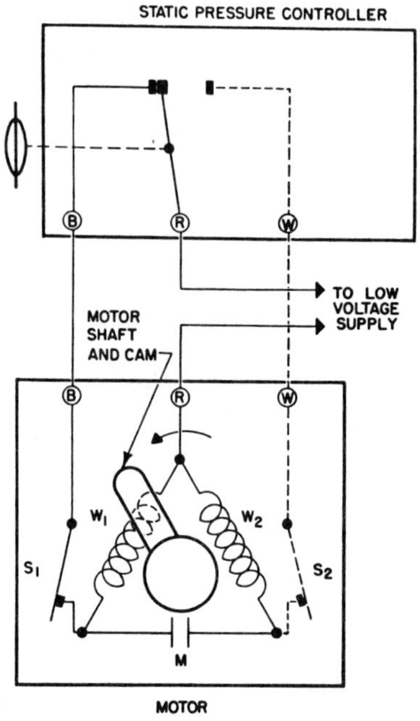

Figure 8-23 Action on a drop in pressure. (*Courtesy of Honeywell Inc.*)

The following is the action when the pressure drop cannot be immediately corrected by the floating action of the controller (see Figure 8-24):

1. The motor continues to drive the final control element toward a corrective position until it reaches its limit of travel.
2. The limit switch S_1 is then opened by the motor cam switch, and the circuit is de-energized.
3. The motor stops turning.

The following is the action on a rise in pressure: When a rise in pressure occurs, the controller blade makes contact with the W terminal. At this time the motor rotates in the opposite direction until the pressure rise has been corrected or until it has reached the limit of its travel path and the limit switch S_2 is opened.

At any time the controller blade is floating between the W and the B terminals, the motor is de-energized. The motor circuit remains de-energized until the pressure moves from between the upper or lower limits of the neutral zone of the controller.

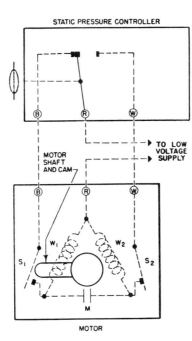

Figure 8-24 Action on a continued pressure drop. (*Courtesy of Honeywell Inc.*)

Series 60 Floating Control Combinations

The Series 60 floating control circuit combinations are not usually very complicated because no limit controls are needed. See Figure 8-25.

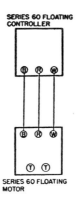

Figure 8-25 Series 60 floating motor-controller combination. (*Courtesy of Honeywell Inc.*)

Series 20 Control Application

The Series 20 control circuit is designed for use on low-voltage, two-position control circuits consisting of

1. Motorized valves

2. Motorized dampers
3. Relays

The Series 20 control circuits are not fail-safe systems and are not to be used where a continued operation of the controlled equipment would be hazardous should there be a power failure. The motors and the equipment under their control will remain in whatever position that they happen to occupy at the time of power failure.

Series 20 control circuits consist of one holding and two starting circuits. The motors used in these systems will operate in only one direction. They make a one-half turn each time either one of the starting circuits and the holding circuit are energized. The holding circuit is made at the beginning and broken at the end of each one-half revolution of the cam and switch arrangement located on the motor armature.

A basic Series 20 control circuit may be built by combining a Series 20 controller with a Series 20 motor and adding the necessary limit controls where they are needed.

Series 20 Control Equipment

The Series 20 control equipment consists of the following devices.

Controllers: The controllers which are used in the Series 20 control circuits are

1. Room thermostats
2. Insertion thermostats
3. Humidity controllers
4. Pressure controllers

The controllers which are designed for use in the Series 20 control circuits include a single-pole double-throw switching action caused by either an open contact or a mercury switch.

The switch may be of the snap-acting type so that the blade will jump from one contact to another or of the slow-acting type so that the blade will float between the two contacts. Regardless of which type is used, one circuit is completed on a rise and the other is completed on a fall in the variable being controlled.

Relays: The operation of a relay from a Series 20 controller is possible. In operation, a universal relay connected to a Series 20 controller is equal to the former Series 30 circuit.

Motor units: The Series 20 motor-type unit can operate either valves or dampers through linkages. Their construction includes

1. A unidirectional electric motor designed to operate on low voltage
2. Gear trains which provide the power at the end of the motor shaft
3. A maintaining switch for the holding circuit

Chap. 8   Basic Electric Control Circuits   143

Being that the Series 20 motor will operate in only one direction, the linkage between it and the damper or the valve must be arranged so that the force exerted by it is in one direction during one-half of the revolution and in the other direction during the other half revolution.

To achieve the two-position operation, the motor and its related circuit are designed so that once the motor is started, it cannot stop until it has rotated a full one-half revolution.

Series 20 Control Operation

A complete Series 20 control circuit consists of the following equipment (see Figure 8-26):

1. An open-contact thermostat
2. A Series 20 motor
3. A step-down transformer

The thermostat has a bimetal blade which makes contact with the *B* contact when a drop in temperature occurs. It also makes contact with the *W* contact on a rise in temperature. See Figure 8-26. In this figure, the thermostat blade

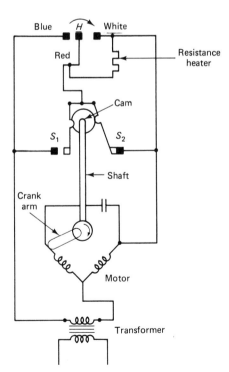

Figure 8-26   A complete basic series 20 control circuit.

is shown in the satisfied position. Also shown are the maintaining switch, the motor, the gear train, and the crank arm in symbolic representation.

In Figures 8-27 through 8-31 the progressive completion of the circuits is shown. In these examples, the cam rotates with the motor armature. Also, in these diagrams, the following lines represent the action that is taking place:

1. A solid black line indicates that a maximum electric current is flowing in that part of the circuit.
2. A broken black line indicates that a negligible amount of electric current is flowing in that part of the circuit.
3. A double black line represents that there is no action in that part of the circuit.

On a drop in the temperature, the following action occurs (see Figure 8-27):

1. The thermostat bimetal blade contacts the *B* contact.
2. The starting circuit is then established.
3. The motor is energized and starts rotating in the clockwise direction.

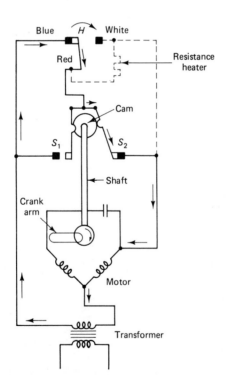
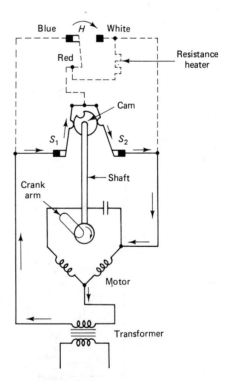

Figure 8-27 Action on a drop in temperature. (*Courtesy of Honeywell Inc.*)

Figure 8-28 The holding circuit is established. (*Courtesy of Honeywell Inc.*)

Chap. 8   Basic Electric Control Circuits

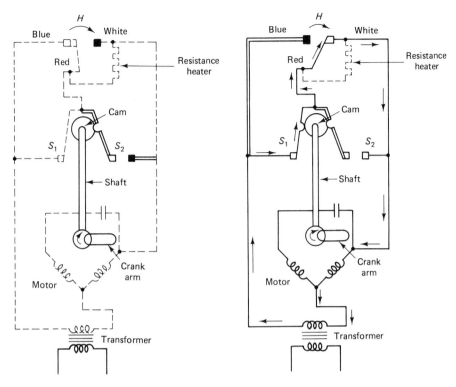

Figure 8-29 The holding circuit is broken. (*Courtesy of Honeywell Inc.*)

Figure 8-30 On a temperature rise. (*Courtesy of Honeywell Inc.*)

The holding circuit is established. See Figure 8-28. As the motor and cam rotate,

1. The blade on the left of the maintaining switch makes contact with S_1, and the holding circuit is established. The holding circuit in these controls is independent of the starting circuit, and once they are made, they will furnish electric current to the motor regardless of the thermostat action.
2. Should the thermostat continue to hold the *B* contact closed, a small amount of electric current will flow through it. Most of the current will, at this time, pass through the holding circuit because it offers the least amount of resistance.

The holding circuit is broken. After the motor shaft has rotated through 180°, the following action occurs (see Figure 8-29):

1. The cam breaks contact S_2.
2. All of the circuits are incomplete.
3. The controller motor stops rotating.

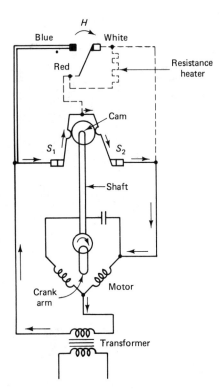

Figure 8-31 The holding circuit is re-established. (*Courtesy of Honeywell Inc.*)

4. A negligible amount of electric current flows through the heater resistance. This amount of current is not enough to cause the motor to rotate.

Both a temperature rise in the controlled space and heat from the resistance heater cause the following actions (see Figure 8-30):

1. The thermostat bimetal blade moves to the right and engages with the W contact.
2. The starting circuit is now completed.
3. The motor starts to rotate in a clockwise direction.

The additional heat from the resistance heater causes the thermostat to close the R to W contacts more quickly than it normally would. This type of action tends to smooth out the overshooting swings which are present in the two-position type of control systems.

When the holding circuit is reestablished, the following action occurs (see Figure 8-31):

1. The blade on the right of the maintaining switch makes contact with S_2.
2. The motor continues its operation until it has rotated one-half turn.

After the motor has completed the one-half turn, the holding circuit is broken again at contact S_1, stopping the motor. The cycle is now complete.

Series 20 Control Combinations

There are several combinations of Series 20 controllers and limit controls. See Figures 8-32 and 8-33.

The circuit in Figure 8-32 is completed between the *B* terminals of the limit control all of the time, with the exception of when a high-limit condition occurs. The electrical connections between the thermostat and the motor generally are straight through, just as if the high-limit control were not in the circuit. In this manner, the thermostat has direct control over the motor except when a high-limit condition exists.

When a high-limit condition exists, the *B* circuit is broken, and the *R* to *W* terminals are made in the limit control. The *R* to *W* connection causes the motor to drive to the closed position even in conditions when the thermostat is calling for heat.

This particular combination of controls is generally used in hand-fired heating applications where the thermostat and the motor work in combination to open and close a draft damper and a check damper. The high-limit control is physically located in the furnace bonnet or boiler and is used to close the draft whenever the bonnet temperature or the water temperature goes above a predetermined high-limit setting.

Low-limit control: A Series 20 three-wire mercury switch is used as the low-limit control. See Figure 8-33. The only time that this control is used is when a low boiler water temperature exists. Otherwise, this control just stands by and is an inactive component in the thermostat to motor circuit because the terminals

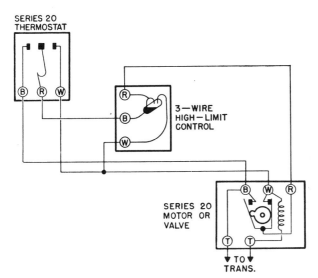

Figure 8-32 A typical series 20 high-limit control circuit. (*Courtesy of Honeywell Inc.*)

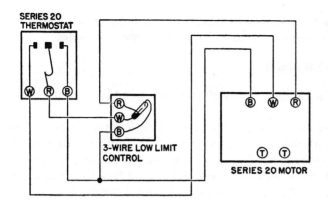

Figure 8-33 A series 20 circuit diagram with a three-wire mercury switch low limit. (*Courtesy of Honeywell Inc.*)

of the thermostat are then connected color to color to the motor terminals.

When a low-limit condition occurs, the mercury bulb will dump in the opposite direction, thus opening the *R* circuit to the themostat and connecting the *R* to *B* terminals at the motor. This action causes the motor to open regardless of the themostat action.

A common Series 20 circuit would consist of a relay with a Series 20 switching action, and a Series 20 manual switch would be in control of a Series 20 motor. See Figure 8-34. This particular combination is commonly used to control a damper on a fan ventilating system.

In this circuit, the relay coil is connected to the motor side of the fan starting switch so that the common and in contacts are closed when the fan is in operation. When in this position the relay puts the manual switch in control of the damper motor. Should the fan be shut down and the operator does not close the damper with the manual switch, the relay will cause it to close automatically when the relay drops out. This action prevents air from entering the ventilating duct when the fan is not running.

Controlling two motors with one controller: This type of Series 20 control circuit generally consists of two Series 20 control motors which are being con-

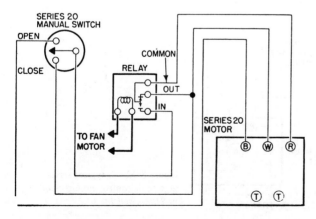

Figure 8-34 A series 20 manual control circuit system. (*Courtesy of Honeywell Inc.*)

Chap. 8   Basic Electric Control Circuits

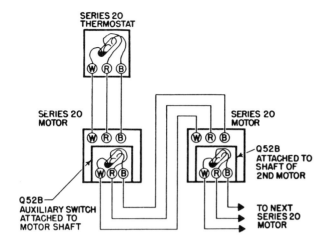

Figure 8-35 A diagram for controlling two motors from one controller. (*Courtesy of Honeywell Inc.*)

trolled by one controller. See Figure 8-35. The control motors cannot be connected in parallel to the same controller because both of the motors would operate as long as the holding circuit of either motor is closed. The successful operation of this type of circuit would require a perfect synchronization of both the motors so that both holding circuits would be broken at the same instant.

It is possible, however, to control two motors as shown in Figure 8-35. The first motor is controlled by the controller directly, and the second motor is controlled by an auxiliary switch which is attached to the shaft of the first motor. As many motors as desired can be connected in this manner, and all can be operated from one controller.

Due to the electrical connections, there is a slight lag between each successive motor. It should be remembered that there is a practical limit to the number of motors that may be controlled in this manner. Usually only five or six will be successful.

Series 20 controller and universal relay: The universal relay and the Series 20 controller are connected together in series to form this particular type of control circuit. See Figure 8-36. When we inspect the diagram, it will be seen that the following functions occur:

1. A drop in temperature at the thermostat will cause the thermostat to short W to B at the relay.
2. A rise in temperature at the thermostat will cause the thermostat to short X to B at the relay.

As an example; Refer to Figures 8-37 and 8-38. In these diagrams the load circuit and the thermostat have been removed for simplification. It should be remembered that the load circuit is closed at any time the relay is pulled in.

When a decrease in temperature at the thermostat occurs, the following things take place:

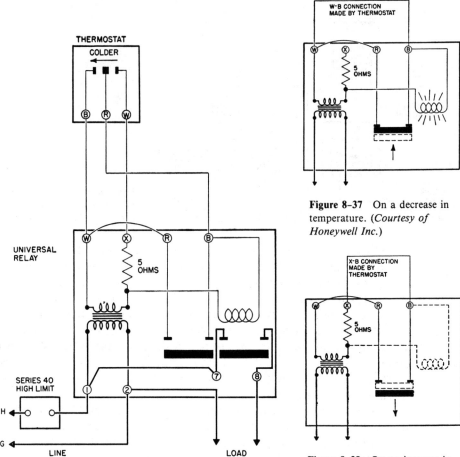

Figure 8-36 Series 20 controller, universal relay, and a high-limit control in series. (*Courtesy of Honeywell Inc.*)

Figure 8-37 On a decrease in temperature. (*Courtesy of Honeywell Inc.*)

Figure 8-38 On an increase in temperature. (*Courtesy of Honeywell Inc.*)

1. As the temperature falls, contacts *W* to *B* are made by the thermostat which close the relay coil circuit.
2. At this time the relay coil is energized and pulls in the relay armature, closing both the holding and the load circuits.
3. The connection between *W* and *B* may now be broken by the thermostat, and the holding circuit will "hold in" the relay.

When an increase in temperature at the thermostat occurs, the following things take place:

1. When the temperature has risen enough, the thermostat makes *X* to *B*.
2. This completes an almost direct short in the path shown by the solid lines

Chap. 8 Basic Electric Control Circuits

in Figure 8-38. The 5-Ω resistor limits the amount of current flowing at this point.
3. A limited power type of transformer is used as the power source, and because of this, there is a critical power drop at the relay coil. (Limited power transformers reduce their voltage on an increase in the load.)
4. The relay then "drops out," breaking both the holding and the load circuits.

It is recommended that a Series 40 limit control be used in these circuits and wired into the circuit as shown in Figure 8-36 at any time a high-limit control is needed.

Series 60 Two-Position Control Application

The Series 60 two-position types of control circuits are very similar to the Series 20 circuit with the exception that the Series 60 is designed for line-voltage operation. They are especially useful for applications such as

1. Industrial applications in which line-voltage equipment is used.
2. Installations which require single-pole double-throw (SPDT) control of the line voltage.

The Series 60 two-position control circuit is not fail-safe in operation and is not recommended where a continued operation of the equipment would present a hazardous situation if the control power should fail. The motors and any equipment under their control will remain in the position that they have when the power failure occurs.

A basic Series 60 control circuit can be formed by combining a Series 60 two-position controller and a Series 60 two-position motor. Any required limit controls can be added when and where required.

Series 60 Two-Position Equipment

The following is a description of the equipment used in Series 60 two-position control systems.

Controllers: The following controllers are designed for use on the Series 60 two-position control circuits:

1. Room thermostats
2. Insertion thermostats
3. Pressure controllers
4. Humidity controllers

Motor power units: The power units which are designed for use on Series 60 two-position control circuits are made up of a small motor which operates directly on line voltage. This motor delivers its power through a gear train to a drive shaft which may be mechanically connected to either dampers or valves.

The result of this type of action is a two-position control. Therefore, it is necessary that the action of the crank arm be divided into two halves of each revolution, between each of which the motor stops rotating. The maintaining switch causes this action. This action is similar to that which is achieved by the Series 20 control units.

Series 60 Two-Position Control Operation

The Series 20 and the Series 60 circuits have almost identical operating characteristics. Therefore, a complete description of the operating characteristics of the Series 60 circuit will not be given at this time. Refer to the Series 20 operation description given earlier.

The mechanical construction of the two different types of equipment is, however, quite different because of the different types of service that is demanded from each of them. Even with these exceptions, their basic designs are almost identical.

Series 60 Two-Position Control Combinations

The different combinations of Series 60 controls are connected identically to the Series 20 control combinations, the one exception being that line-voltage power rather than low-voltage power is used to power the circuit. The controls must either be the mercury bulb type or have snap-acting contacts incorporated into them to prevent arcing between the contacts when switching line-voltage loads.

Series 90 Control Application

The Series 90 control circuit is designed to provide modulating operation or proportioning control action and is applicable to such devices as

1. Motorized valves
2. Motorized dampers
3. Sequencing switching mechanisms

The Series 90 control circuit is designed to operate and position the controlled device at any point between full open and full closed which will proportion the delivered medium to the exact need as indicated by the controlling device.

Modulating-type control circuits are not hampered by the same limitations as are two-position or floating-type control circuits.

As an example,

1. When modulating-type control circuits are used, the power unit, when energized, will operate only long enough to move the final control element

a distance which is proportional to the change in the controlled variable. A Series 90 thermostat, therefore, with a 2° proportional band will move the actuator one-twentieth of its total travel for each 1/10° change in the temperature. Series 90 power units will move a definite amount for every change in the controller position.

2. Once a two-position power unit is energized, it is limited in its operation by the action of the maintaining switch. The motor must advance to one of its extreme positions and remain there until the conditions at the controller have changed through the entire range of its differential.
3. Once the floating control power unit is energized, it is limited in its operation by the amount of time required to have the change in its position reflected at the controller location.

To form a complete Series 90 control circuit, combine any Series 90 controller with a motor power unit of a relay which is constructed for providing the proportioning action. The required limit controls may be added where desired. Automatic compensation may also be provided when desirable.

Series 90 Control Equipment

The following is a description of the Series 90 control circuit equipment.
Controllers: The following is a description of the controllers used on Series 90 control circuits:

1. Room thermostats
2. Insertion thermostats
3. Humidity controllers
4. Pressure controllers

The Series 90 controllers are different from those of the other types of series. In this type of controller, the electrical mechanism is a variable potentiometer rather than an electric switch. It is equipped with a contact blade that moves across the potentiometer winding. This winding has exactly 135-Ω of resistance. The contact blade is moved in response to temperature, pressure, or humidity changes in the space being controlled.

Motors: The following is a description of the Series 90 motors. These motors consist of the following:

1. A reversible capacitor power unit
2. A balancing relay
3. A balancing potentiometer
4. A gear train

The motor power unit consists of a low-voltage capacitor-type motor which

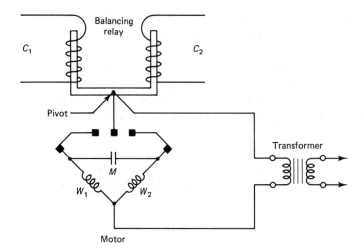

Figure 8-39 Balancing relay and a series 90 motor diagram. (*Courtesy of Honeywell Inc.*)

turns the motor drive shaft through the use of a gear train. There are limit switches which are operated by the motor to limit the rotation to 160°. All the moving parts, including the gear train, are immersed in oil to eliminate the need for periodic lubrication and to ensure a long, quiet life.

The power unit operation is started, stopped, and reversed by a single-pole double-throw set of contacts in the balancing relay. See Figure 8-39. The balancing relay incorporates two solenoid coils with parallel axes. The legs of the U-shaped armature are inserted into the holes inside the coils. The armature has a pivot at the center so that it can be tilted by the changing magnetic flux in the two coils.

The armature has a contact arm fastened to it so that it will touch one or the other of the two stationary contacts as the armature is moved back and forth on the pivot. At any time the relay is balanced, the contact arm floats between the two stationary contacts, not touching either of them.

The motor has a balancing potentiometer which is included as an integral part. This potentiometer is identical to the one in the Series 90 controller. It is made up of a 135-Ω coil of wire and a movable contact arm. The arm is moved by the motor shaft so that it travels along the coil. Contact is established where it touches the coil. See Figure 8-40.

Series 90 Control Operation

The following is a description of the operation of a Series 90 control circuit.

Balancing relay operation: The balancing relay is made of two coils which are equipped with a common armature mounted on a pivot. See Figure 8-39. The relay, as it is applied to a Series 90 control circuit, has electric current flowing through the coils. The amount of current flowing is regulated by the relative positions of the controller potentiometer and the motor balancing potentiometer. Therefore, when there are equal amounts of current flowing through both of the balancing relay coils, the contact blade is located in the center of the space

Chap. 8 Basic Electric Control Circuits

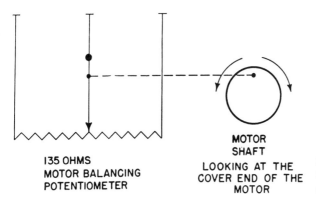

Figure 8-40 Diagram of series 90 balancing potentiometer. (*Courtesy of Honeywell Inc.*)

between the two stationary contacts, and the motor is not turning in any direction. As the finger of the potentiometer inside the controller is moved in response to a change in the controlled variable, a greater amount of current flows through one of these coils than flows through the other coil, and the relay becomes unbalanced. The relay armature is then rotated so that the blade touches one of the stationary contacts, and the motor turns in the corresponding direction. See Figure 8-41.

Referring to Figure 8-39, if the relay coil C_1 should receive more of the

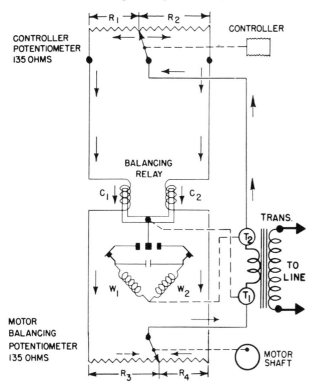

Figure 8-41 A series 90 control circuit in the balanced condition. (*Courtesy of Honeywell Inc.*)

current than coil C_2 and has a stronger magnetic field, the contact blade will move to the left, and the circuit between the transformer and the motor winding W_1 is completed. The electric current also flows through the capacitor M to the motor winding W_2. The motor will turn in the corresponding direction until the balancing relay is balanced and the circuit is broken.

Should the coil C_2 receive more of the current than coil C_1, the circuit is completed directly to the motor winding W_2, and the motor will turn in the opposite direction.

Motor balancing potentiometer: The contact made by the balancing relay can be broken only if the amount of current flowing through coil C_1 becomes equal to the current flowing through coil C_2. This action is caused by the motor balancing potentiometer which is linked to the motor shaft. When the motor turns, the finger of the motor balancing relay is driven toward a position which will equalize the resistances in both legs of the control circuit.

For each position of the motor shaft there is a corresponding position of the contact finger for each degree of rotation through its complete 160° of travel. As an example, when the motor shaft reaches 40° from one of its ends of travel, 25% of its arc, the contact finger is at $33\frac{3}{4}$ Ω, a value that lies 25% of the distance from the corresponding extreme of the coil resistance.

The complete sequence of operation of the Series 90 control circuit is shown in Figures 8-41, 8-42, and 8-43.

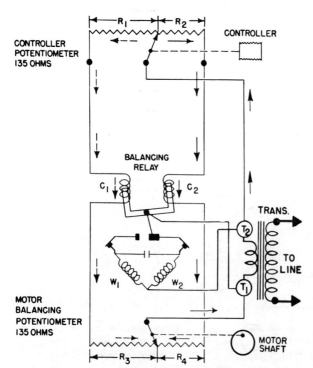

Figure 8-42 A series 90 control circuit on a drop in temperature. (*Courtesy of Honeywell Inc.*)

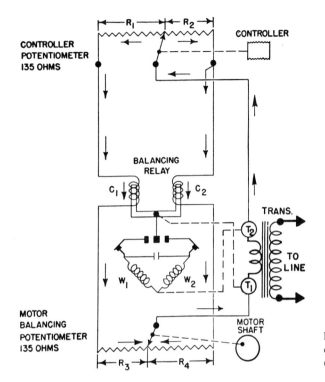

Figure 8-43 A series 90 control circuit rebalanced in a new position. (*Courtesy of Honeywell Inc.*)

As shown in Figure 8-41, there is an instantaneous condition in which the current is flowing from the transformer, through the potentiometer finger, and down through both legs of the control circuit. In the positions shown, the thermostat potentiometer finger and the motor balancing potentiometer finger divide their respective coils so that $R_1 = R_4$ and $R_2 = R_3$. Thus, $R_1 + R_3 = R_2 + R_4$, and the resistances on both sides of the circuit are equal. The coils C_1 and C_2 in the balancing relay are equally energized, and the armature of the balancing relay is balanced. The contact arm is floating between the two stationary contacts, no current is flowing to the motor, and the motor is not operating.

As shown in Figure 8-42, another instantaneous condition exists in which the temperature has dropped a small amount. Because of this drop, the thermostat potentiometer finger has moved toward the right-hand end of the potentiometer coil. The amounts of resistance on both sides of the circuit are no longer equal. $R_1 + R_3$ is greater than $R_2 + R_4$. The larger part of the current is now flowing through the right-hand leg of the circuit, and coil C_2 of the balancing relay puts more force on the armature than C_1 does. The armature has rotated, making contact with the side of the circuit that directs the current to the motor winding W_2. The motor then turns in the corresponding direction and moves the motor balancing potentiometer to a new position.

As shown in Figure 8-43, an instantaneous condition exists after the motor shaft has moved the balancing potentiometer to a position which equalizes the current flow through the two legs of the circuit. In this condition, the right-hand side of the thermostat potentiometer has a resistance which equals that of the left-hand side of the motor balancing potentiometer. Again, $R_1 + R_3 = R_2 + R_4$, the current flowing through the two legs is equal, and the motor is not turning.

If we carefully analyze these diagrams, it should be seen that the motor turns until the contact finger of the motor balancing potentiometer reaches a position which corresponds to the position of the finger of the thermostat potentiometer.

Series 90 Control Combinations

The following is a description of the Series 90 control combinations.

Low-limit control: The low-limit type of control is commonly used in Series 90 control circuits which are applied to central fan ventilating or air conditioning systems in which the means of heating is controlled from a room thermostat or a return air duct controller. The space temperature can rise rapidly as a result of increased solar radiation, increased occupancy, or any other type of condition which would result in a sudden decrease in the heating load. The room or the return air controller will then be satisfied and will close off the final control element.

Should the system be one that takes in a portion of outside air, it is quite possible that the air could be discharged into the room at too low a temperature. When this supply air temperature drops below about 60°F, the room will probably feel drafty.

The installation of a low-limit control in the return air duct is often a desirable part of the control system. This controller regulates the coil valve to maintain the discharge air temperature at a comfortable level and thus limit the lower temperature of the discharge air.

It should be noted that the low-limit controller should be set at the lowest temperature that can possibly be supplied without drafts.

In most cases, the low-limit controller has a potentiometer resistance of 135Ω. This permits the controller to operate the final control element over 50% of its operating range. Controllers which are equipped with double potentiometers can be used for providing the final control with 100% of its operating range. These potentiometers must be connected in series so that a resistance of 270Ω can be obtained. In most cases, however, the 50% range is more desirable. Therefore, the discussion that follows makes reference to the limit control which is equipped with the 135Ω potentiometer.

Figure 8-44 shows the external wiring connections for a Series 90 room thermostat, low-limit control, and motorized valve.

A typical Series 90 circuit with a low-limit control is shown in Figure 8-45. This diagram shows the low-limit control being satisfied, and the potentiometer

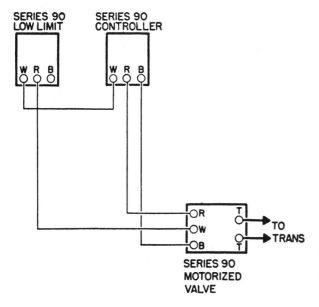

Figure 8-44 Series 90 motorized valve, thermostat, and low-limit control diagram. (*Courtesy of Honeywell Inc.*)

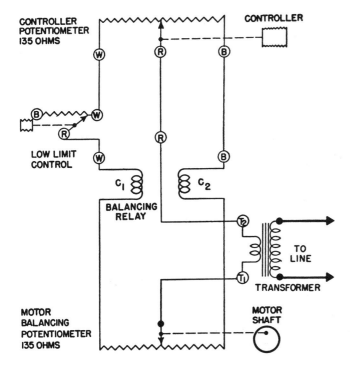

Figure 8-45 Series 90 circuit with low-limit control satisfied. (*Courtesy of Honeywell Inc.*)

finger is on the *W* end of the winding. This position is indicative that the temperature of the discharge air is above the setting of the low-limit control. When this condition occurs, the controller operates the valve motor in exactly the same way as in the Series 90 circuit without the low-limit control described in the section "Series 90 Control Operation." As long as the low-limit controller remains satisfied, its potentiometer winding will have no effect on whether or not the relay is balanced. The relay coils are affected only by the controller potentiometer and the motor balancing potentiometer.

A typical Series 90 circuit with a low-limit control in which the room temperature is beyond the modulating range of the room thermostat and the low-limit control is in control of the valve motor is shown in Figure 8-46. In this figure the finger of the room controller potentiometer has moved to the *W* end of the coil. The low-limit control remains satisfied, and the motor is in the closed position of the motor balancing potentiometer.

Should the system begin taking in outside air, the discharge air temperature may drop. As this temperature drops into the modulating range of the low-limit control and approaches its setting, the potentiometer finger moves away from the *W* end of the coil to a new position; see the dashed lines.

If we assume that the finger has moved to a new position which is equal

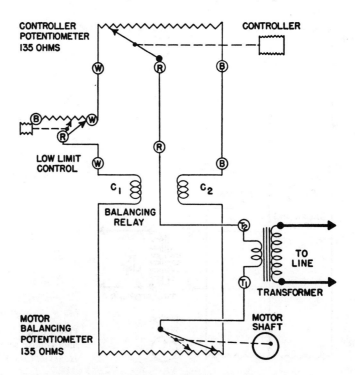

Figure 8-46 Series 90 circuit with low-limit controlling the valve or motor. (*Courtesy of Honeywell Inc.*)

to 25% of the total travel across the potentiometer coil, 25% of the 135Ω, or $33\frac{3}{4}$ Ω, of resistance is introduced into the left leg of the control circuit. The total circuit then becomes unbalanced, with most of the current flowing through the coil C_2 of the balancing relay. The armature of the balancing relay is positioned so that it makes contact, and the motor balancing potentiometer finger is moved toward the open end of the coil as the valve is opened. When the potentiometer arm has moved far enough to cause the resistances of the circuit to be equal again, the circuit is broken by the balancing relay, and the motor stops.

High-limit control: The high-limit control generally has a 135 Ω potentiometer incorporated into it to obtain better results. However, where a 100% operating range is desired, high-limit controls are equipped with dual potentiometers.

The external wiring diagram for a Series 90 high-limit controller and a motorized valve is shown in Figure 8-47. This type of circuit is used when there is danger of the temperature rising too high. The high-limit control will take control of the motorized valve whenever the discharge air temperature enters into its modulating range.

A complete Series 90 control circuit with a high-limit controller is shown in Figure 8-48. This circuit is similar to the low-limit control circuit which is shown in Figure 8-46, with the exception that the high-limit control is connected into the *B* leg of the control circuit. The operation of the high-limit control is just like the operation of the low-limit control.

When the temperature of the discharge air rises into the modulating range of the high-limit controller, the finger starts moving toward the *W* side. This results in more resistance being introduced into the *B* leg of the circuit, and the

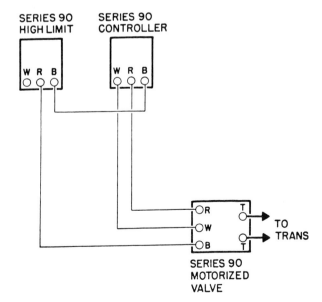

Figure 8-47 Wiring diagram for a series 90 circuit with a high-limit control. (*Courtesy of Honeywell Inc.*)

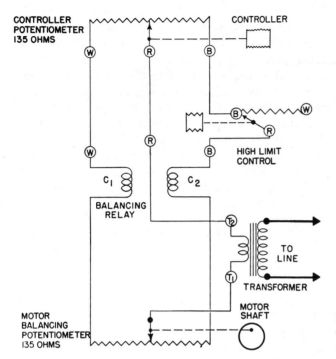

Figure 8-48 A complete series 90 circuit with a high-limit control. (*Courtesy of Honeywell Inc.*)

relay coil C_1 becomes a stronger magnet than the coil C_2. When contact is made by the relay armature, the motor turns toward the closed position until the motor balancing potentiometer reaches a final position in which the resistances of each of the legs of the circuit are again equal.

Two-position limit controls: In Series 90 control circuits it is sometimes desirable to use two-position limit controls in applications when it is unnecessary or undesirable that the limit control operate on a proportional action basis. This type of two-position limit control must always be of the snap-acting, single-pole double-throw type. Two-position limit controls must not be used in applications where the temperature of the medium in which their sensing elements are located is affected to any great extent by the opening and closing of the valve or the damper which is under its control. As an example, let us suppose that a Series 60 high- or low-limit control is used in a Series 90 control circuit as a limit control in a discharge air duct. At any time the temperature of the air enters the operating range of the limit control, the valve controlling the steam to the heating coil will cycle on and off continuously.

Two-position-type limit controls are sometimes used in applications which require humidity high-limit control in a cooling application. This circuit uses a single-pole double-throw switching action. See Figure 8-49. When the humidity drops below the range of the humidity high-limit controller, the circuit is completed from the thermostat to the motor, and the cooling valve is controlled in

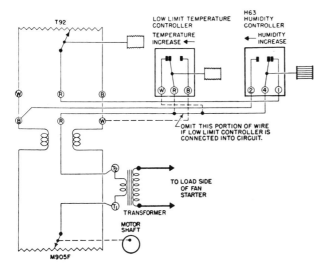

Figure 8-49 Series 90 circuit with a two-position humidity high-limit control. (*Courtesy of Honeywell Inc.*)

a normal fashion. Should the humidity rise into the range of the high-limit control, the limit control shorts terminals *R* to *B* at the motor, and the motor then opens the cooling valve to the maximum open position. When a power failure occurs, or when the fan is shut down, the cooling valve is closed. Some type of reheat device should be installed with a system of this type, because the lowering of the air temperature will result in a higher humidity condition. To successfully remove both the sensible and the latent heat, the air should be cooled down to the dew point and then be reheated back up to the desired temperature.

When a reheat system is not included in the installation, a second temperature controller should be used to act as a low-limit control. This is especially applicable in highly humid climates.

The low-limit control should be wired into the circuit as shown in Figure 8-49, and it should be set a few degrees below the main temperature controller. In the event the air temperature should fall too low, the low-limit controller will take control away from the humidity control, and the cooling valve will close.

Manual and automatic switching: The use of manual switches and relays in Series 90 control circuits to perform certain functions is often desirable. Manual switches are illustrated in the diagrams throughout the remainder of this chapter. Relays, however, that have the same switching action as the manual switches are often used where automatic switching is desired.

Figure 8-50 illustrates a manual switch with a single-pole double-throw switching action as used in a Series 90 control circuit. When the switch is placed in the automatic position, as shown in the figure, the *R* circuit to the motor is completed, and the motor turns in a normal fashion under the control of the controller. When the switch is placed in the closed position, the *R* wire from the controller is broken, and the *R* to *W* wires are completed at the motor. The motor then advances to the closed position. This type of system is often used in fan

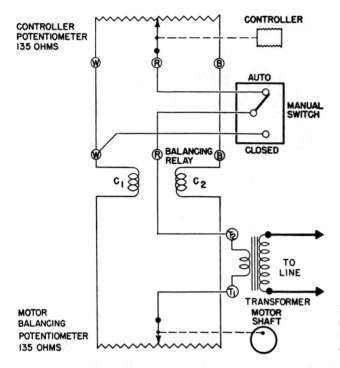

Figure 8-50 Series 90 circuit with single-pole double-throw manual switch for automatic or closed operation. (*Courtesy of Honeywell Inc.*)

heating systems to manually shut off a part of the controls when it is not required that they operate.

Transfer of motor from one thermostat to another: A double-pole double-throw switch can be used to transfer the control of a single Series 90 motor from one thermostat to another. To prevent interaction of the two thermostats, both the *W* and the *B* wires must be broken when the Series 90 thermostat is taken out of control. The *R* wire does not have to be broken. See Figure 8-51.

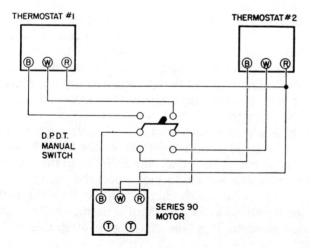

Figure 8-51 Diagram of circuit for transferring control of a series 90 motor from one thermostat to another. (*Courtesy of Honeywell Inc.*)

Chap. 8   Basic Electric Control Circuits

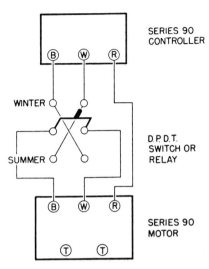

Figure 8-52 Wiring diagram arrangement used for reversing control so that the same thermostat and final control element can be used for both heating and cooling. (*Courtesy of Honeywell Inc.*)

Reversing for heating and cooling control: In applications where it is necessary to use the same thermostat and the same final control element for both the heating and the cooling, the wiring can be done as shown in Figure 8-52. The double-pole double-throw manual switch is used for reversing the wiring from the thermostat to the motor for the cooling control so that the motor will operate in the opposite direction than when used for heating control. When the switch is placed in the *winter* position, the thermostat and the motor are connected color to color for heating. When the manual switch is placed in the *summer* position, the connections are *B* to *W* for cooling.

Transfer of thermostat from one motor to another: In some installations it is desirable to operate two Series 90 motors, one at a time, from one single controller. See Figure 8-53. A triple-pole double-throw manual switch is used

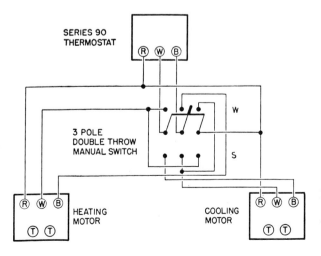

Figure 8-53 Diagram of arrangement for transferring thermostat to control one or the other of two motors. (*Courtesy of Honeywell Inc.*)

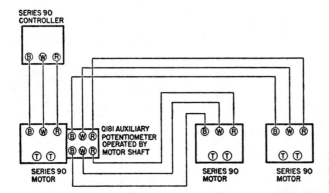

Figure 8-54 Diagram for controlling motors in sequence from a single controller. (*Courtesy of Honeywell Inc.*)

to accomplish the switching action. When the switch is placed in the summer position, the control circuit is completed from the thermostat to the cooling motor with the *B* to *W* and *W* to *B* wires, and the motor then operates normally as under the control of the thermostat. The manual switch, at the same time, connects *R* to *W* at the heating motor, making it positively closed.

When the manual switch is placed in the winter position, the thermostat is connected color to color with the heating control motor which operates normally in this condition. The cooling motor is positively closed when the circuit is in this condition.

Sequence control: A good method of controlling Series 90 motors in sequence from one thermostat is shown in Figure 8-54. The Q181 auxiliary controller which contains two potentiometers is fastened to the shaft of the master control motor. Each one of the potentiometers has control over one of the auxiliary motors. Therefore, the controller is in direct control of the master motor. The position of the motor shaft regulates the two potentiometers, which in turn regulate the auxiliary motors.

The starting points and the proportional bands of the two auxiliary potentiometers can be adjusted to provide the desired sequence of operation.

Unison control: One method of attaining unison control of two motors by one controller is shown in Figure 8-55. The master motor is a Series 90 motor with a built-in dual potentiometer. The dual potentiometer operates the second motor. A third motor can also be added by installing more motors with dual potentiometers in the first and the second positions and controlling the third motor from the dual potentiometer in the second motor. Three motors are the maximum that should be controlled in this manner.

Another method that may be used for controlling several motors in unison is shown in Figure 8-56. The Q68B auxiliary potentiometer is mounted on one end of the master motor shaft. Q68 potentiometers are available for controlling up to eight motors in parallel. It must be remembered that the Q68 potentiometer must be used to control the specific number of motors being used.

Manual minimum positioning: A typical outdoor air control system using a manual potentiometer for minimum positioning is shown in Figure 8-57. When

Chap. 8   Basic Electric Control Circuits   167

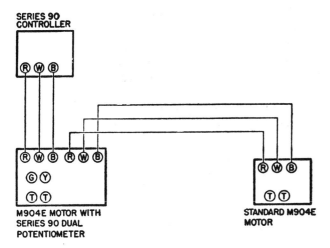

Figure 8-55 Diagram for unison control of two motors from one controller, using a master motor with built-in dual potentiometer. (*Courtesy of Honeywell Inc.*)

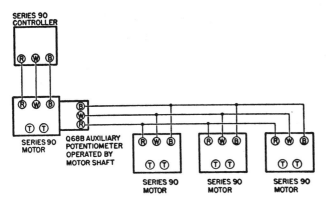

Figure 8-56 Diagram for unison control of three auxiliary motors and one master motor by means of an auxiliary potentiometer. (*Courtesy of Honeywell Inc.*)

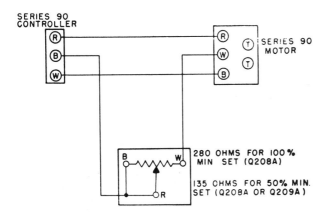

Figure 8-57 Diagram for manual minimum positioning of typical outdoor air control system. (*Courtesy of Honeywell Inc.*)

the pointer is turned so that the pointer is at *W*, the potentiometer coil is shorted out of the circuit, and the motor will operate normally, with the damper going completely closed should the controller demand it. If the pointer is turned so that the wiper is toward *B*, the resistance between *B* on the controller and *W* on the motor is increased, and travel of the motor toward the closed position is limited.

A manual potentiometer with 135Ω of resistance will provide up to 50% of the minimum position opening, and a 280Ω potentiometer will provide up to 100% of the minimum position opening.

Recycling step controllers (Honeywell): The wiring diagram of a Series 90 step controller used to control five compressors in a large fan cooling system is shown in Figure 8-58.

Step controllers are available with from 5 to 10 mercury switches operated in sequence by means of a cam assembly which is turned by the control motor. The mercury switches are wired into the compressor starter circuits, and as the controller calls for more cooling, the Series 90 motor rotates, turning on more compressors one at a time as needed. Thus, the correct amount of refrigeration for the load condition is always available.

When there is a power failure or if the system is shut down, it is desirable that the system recycle to the starting point so that the compressors will start one at a time and not overload the electrical circuit.

In the arrangement shown in Figure 8-58, the control motor remains in the position that it occupied when the current went off. However, before any of the compressors can operate again when the power is restored, the control motor

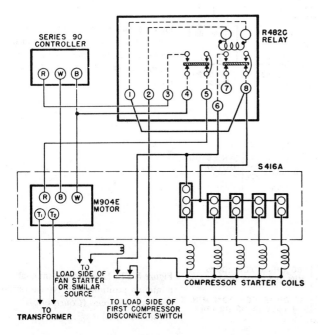

Figure 8-58 Wiring diagram of a Honeywell recycling step control system. (*Courtesy of Honeywell Inc.*)

Chap. 8 Basic Electric Control Circuits   169

must first travel back to the closed position. It then travels back to the position called for by the thermostat, activating the required number of compressors as it returns.

The complete compressor wiring diagram is not shown in the figure to simplify the diagram. The power supply for the compressor coils is often taken from the fan circuit, thus making certain that the compressors will not run when the fan is shut down.

This is valuable in multiple-element control installations because it provides safe, economical compressor operation for all operating conditions with additional assurance that the starting loads will never be great enough to overload the power line.

REVIEW QUESTIONS

1. What is considered to be line voltage in control systems?
2. What is the name for a resistor with a wiper blade that is used in control systems?
3. What are the contacts called that close when the relay coil is de-energized?
4. When using additional listed controls, is the basic circuit changed?
5. What is the switching action of a Series 40 controller?
6. What is done when the current load exceeds the current rating of the controller?
7. What makes a Series 40 controller fail-safe?
8. In the Series 40 high-limit control, why must the thermostat and high limit be snap acting?
9. What are the most popular Series 80 actuators?
10. Are Series 80 controllers designed to switch line voltage?
11. Is the Series 10 control system a modulating or a two-position circuit?
12. Which of the series circuits operate with two bimetal blades?
13. In the Series 10 control circuit, what circuit does the *B* terminal control?
14. How many devices can be controlled at any one time from a single Series 10 controller?
15. On what type of voltage is a Series 60 circuit designed to operate?
16. To what type of device are Series 60 controllers not generally applied?
17. Can snap-acting controllers be used in Series 60 circuits?
18. What do the limit switches inside a motor do?
19. When is the motor in a Series 60 control circuit de-energized?
20. Are the Series 20 control circuits fail-safe?
21. What is the purpose of a maintaining switch in a Series 20 circuit?
22. Will the Series 20 motor operate in both directions?
23. When is a Series 20 low-limit control used?
24. What is the maximum number of motors that can be controlled with a Series 20 controller?
25. What do limited power transformers do on an increase in the load?
26. What is the main difference between the Series 20 and the Series 60 two-position circuits?

27. What type of action does the Series 90 control circuit produce?
28. What type of electrical mechanism is used in the Series 90 controllers?
29. What is the resistance of a potentiometer?
30. Of what does the Series 90 motor power unit consist?
31. What causes the balancing relay to operate?
32. What type of potentiometer is in the Series 90 motor?
33. What regulates the amount of current flowing through the balancing relay coils?
34. At what temperature is the low-limit controller set in a Series 90 control circuit?
35. What could be done to the controller so that it can operate the final control element over 100% of its operating range?

Index

A

Actuators, 18, 117
Air motion relay, purpose, 90
Automatic control system, 12
    basic functions of its parts, 15
Average value of electrical current, 99
Averaging relay, use, 91

B

Balancing relay operation, 122
Basic dual-pressure pneumatic system, 36–38
Basic pneumatic control system, 26–30
Building zoning, 3

C

Calculating current in dc circuits, 100
Control element, final, 6 (*see also* final control element)
Controlled system, 12
    electric, 1
    pneumatic, 2
    self-contained, 1
Controlled variable, 13
Controller, 5, 15
    mechanisms, 17
Control modes, 7, 8, 9, 18
Control theory, 2
Control valve motor, 3
Controlled variable, 13
Controller, 5, 15
    mechanisms, 17
Corrective action, 14
Cycling, 14

D

Day/night system, 39 (*see also* Basic dual-pressure pneumatic system)
Desired value, 13
Deviation, 13
Differential gap, 14
Diodes and rectifiers, use, 105
Direct-acting controller, 31
Disturbance sensing element, 5
Diverting relays, use, 83

E

Effective value of electrical current, 99
Electric circuit components, 95
Electric-pneumatic relay application, 86
Equipment, power rating, 102

F

Final control element, 6, 14 (*see also* Control element, final)
Floating control, 8, 23 (*see also* Control modes)

G

Gradual switch operation, 92

H

Heating effect of electrical current, 99

I

Impedance, 103
 in a bridge circuit, 109
Inductance, 102
Instantaneous value of electrical current, 99

L

Lag, 14, 23

M

Manipulated variable, 13
Master/submaster controllers, 46–49
 application, 46–49
Maximum value of electrical current, 99
Methods of transmitting energy to an actuator, 18
Minimum positioning switch, use, 92
Modes of automatic control, 18
Multiposition (multistage) control, 8 (*see also* Control modes)

O

Offset, 14
Ohm's law, use, 95–100
On-off (two-position) control, 7 (*see also* Control modes, 18)

Index

P

Parallel circuit characteristics, 97
Pneumatic control system components, 31
Pneumatic control valve classifications, 60
Pneumatic control valve components, 59
   actuator, 59
   disc, 59
   guide, 59
   port, 59
   trim, 60
Pneumatic relay operation, 85
Pneumatic switch types, 91
Pneumatic thermostat calibration, 32
Potentiometer connections, 122
Pressure controllers, 44–46
Pressure selector relay application, 89
Primary element, 14
   humidity sensing, 17
   pressure sensing, 17
   temperature sensing, 16
Proportional band, 14
Proportional control, 22
Proportional-plus-reset control, 23
Proportioning control, complex variations, 10
Proportioning (modulating) control, 9
   (*see also* control modes)

R

Receiver controllers and transmission systems, 49–57
Reset chart, 49
Reset percentage, determining, 55
Reset range, 48
Reversing relays, use, 84

S

Sensors, 110–12
Series circuit characteristics, 96
Series 40, controllers, 126
   motor units, 126
   relays, 126
   solenoid valves, 126
   unit heater control, 127
Set point, 5, 13
Silicon controlled rectifiers, 107
Simple control system example, 6
Single-pressure thermostat, 32, 35
Steam to hot water converter selection, 68
Steam valve selection, 66, 70
Summer-winter pneumatic system, 36
   (*see also* Basic dual-pressure pneumatic system)

T

Temperature controllers, 41–44
Thermostat calibration, 33
Thermostats, 1, 2, 3
Throttling range, determining, 35
Transistors, 108
Transmitter sensitivity, 51
Two-position control, 18, 19, 20 (*see also* Modes of automatic control)

V

Valve, capacity index, 63
   spring ranges, 65
Valve piping arrangements, 67

W

Water circulation system, purpose, 77, 79–82
Water valve selection, 73, 77

Z

Zener diodes, 106

TH
7687.5
.L364
1985

14.95

Southwest Campus
St Philip's College

TH
7687.5
.L364

1985